# AGRICULTURAL POLICY IMPLEMENTATION: A CASE STUDY from WESTERN KENYA

The Project Area

Women's Roles & Gender Differences In Development.

# AGRICULTURAL POLICY IMPLEMENTATION: A CASE STUDY from WESTERN KENYA

Kathleen Staudt

KUMARIAN PRESS
West Hartford

Printed in the United States of America

Cover design by

Timothy J. Gothers

Library of Congress Cataloging in Publication Data
Staudt, Kathleen A.
Agricultural policy implementation.
Bibliography: p. 65
1. Agriculture and state—Kenya—Case studies.
2. Women farmers—Kenya—Case studies. 3. Sex discrimination against women—Kenya—Case studies.
I. Title.
HD2126.5.Z8S73 1985 338.1'86762 84-29737
ISBN:0-931816-18-1

KUMARIAN PRESS
29 Bishop Road
West Hartford, CT 06119

# CONTENTS

# LIST OF ILLUSTRATIONS

TABLES

BOXES

# PREFACE

Why should development planners and scholars of development be concerned about women's roles and gender differences?

No project that expresses its goals in terms of production gains or increased benefits can afford to ignore the economic potential and needs of one-half of the population. Guidelines for the design and evaluation of development projects sensitive to women's roles have often been applied only to a narrow range of "women's projects." Our view at the Population Council is that all development efforts could be improved if the differential impact on both class and gender groupings were considered.

The series of case studies on Women's Roles and Gender Differences in Development was developed to demonstrate that such analyses are not only essential, but also feasible within existing structures.

These case studies make clear how inattention to women's roles and gender differences is played out as projects are implemented. Excluding gender as a variable, or limiting women's roles to the welfare sector, results in unintended effects, sometimes positive, but more frequently negative. Many of the stated objectives of the development schemes under study were not attained because project designs were predicated on an incomplete picture of the society to be served and drawn into participation.

The case studies draw largely from material that existed originally in other forms (such as exceptional Ph.D dissertations). From these materials has been extracted the "case:" (1) salient aspects of the culture and society in which the development project was placed, (2) the project dynamics themselves, and finally, (3) an assessment of gains and losses in different goal areas. To complement individual case studies, the Series for Planners includes monographs on broader development phenomena whose effects are seen outside the confines of specific development schemes. As of this writing, the Series includes two monographs, one dealing with the effects of male out-migration on rural women's roles and a second on the impact of different styles of agrarian reform on women's roles and productivity.

These materials are intended to be used by students of development and professionals in the field, including those at the highest planning levels. By providing examples of how individual development schemes have operated vis-a-vis gender, we hope they stimulate in the reader an interest in exploring what these effects might be in development projects being designed, implemented, or evaluated. For some time now, an understanding of class dynamics has been seen as essential in designing projects for successful outcomes. We have the same conviction regarding the importance of understanding gender differentials. We hope that this study series positively advances that notion and provides its readers with new skills and insights by raising questions and suggesting alternatives.

We wish to thank each of our individual authors for the exhaustive work they have put into forming their material into case studies. We commend Marilyn Kohn for her fine editorial work in finalizing the material.

*Judith Bruce*
*Associate*
*The Population Council*

*Ingrid Palmer*
*Editor of the Series*

# SUMMARY


## BACKGROUND: POLICY OBJECTIVES, RURAL ECONOMY, AND RESEARCH DESIGN

Many Kenyans depend on agriculture for their livelihood. In its efforts to increase productivity, improve farmers' livelihood, and expand export crops, the Kenyan government relies on an agricultural extension service to provide information to farmers.

In a large part of western Kenya, men have traditionally migrated elsewhere in search of wage labor, while women stayed at home and managed agricultural production. Even prior to the establishment of this pattern, women performed a major portion of the agricultural labor and management of the food supply. Virtually all women belong to organizational networks through which they share agricultural labor and provide mutual aid. Men dominate administrative offices and political authority networks which provide contacts and information about valuable agricultural services.

The purpose of this study is to compare the access of female-managed farms to agricultural services with that of jointly managed farms (i) to determine whether there are inequities and (ii) to evaluate the consequences of those inequities for female managers' farm performance and, by extension, overall agricultural performance in Kenya.

Seven agricultural policy objectives derived from official documents and implementation realities in the Kakamega District of western Kenya frame this analysis, as summarized below:

1. To assist all farmers to improve agriculture through a field-oriented extension service;

2. To provide specialized knowledge to farmers through formal training programs on an equitable basis;

3. To provide production loans to farmers on an equitable basis;

    a. To encourage cooperative membership and receipt of cooperative loans;

4. To provide available technology to cooperative members in an equitable manner;

5. To provide other cooperative benefits to members who apply for them;

6. To supply information to all farmers about hybrid maize and cash crops; and,

7. To decrease farmers' risk through adaptiveness and diversification.

The study is based on field research carried out in western Kenya from 1974-1975 in three types of administrative settings: ordinary, intensified, and saturated. Survey data from 212 households in Shikulu the ordinary sublocation representing a standard or typical level of services, are first analyzed to identify typical patterns of service for female and jointly managed farms. Then the study examines access in Shitoli the "intensified" sublocation where, as the term suggests, increased services are provided, to assess whether intensifying staff and resources affects disparities between household types. Finally, a subsample in Shitoli of the Lirhembe multipurpose cooperative members is analyzed to determine the effects of further increasing or "saturating" services. A quasi-experimental design is used to assess the impact of inequitable service delivery. Two administrative units, similar in culture, economy, and geography, but different in administrative intervention, allow qualified judgments to be made about programmatic causes of differential performance.

Households, the units of analysis in this study, are divided into female-and jointly managed farms. The women in female-managed farms, two-fifths of the sample, were either widowed or separated from their husbands, who had migrated outside the area for wage employment. Jointly managed farms had a man present, but they have not been termed "male-managed" because of women's substantial involvement in farm work and decision making, even with husbands present. These jointly managed farms are husband-wife teams, or corporately managed farms with intergenerational management since land has not yet been parcelled to sons. Data show that female- and jointly managed farms are similar in their economic standing and in the amount of land they farm.

## FINDINGS

The analysis reveals that the objectives outlined above were not met, due to the preferential delivery of services to farms with a man present. In sum, female-managed farms always received fewer services than jointly managed farms, and gaps increased as the services become more valuable. Comparisons of both sublocations show that female managers' access to services in intensified areas was greater than in ordinary areas, but that gaps between management types remained the same. While gaps between the sexes were smallest under conditions of saturated services in the multipurpose cooperative sample, except in the allocation of valuable grade cows, women managers were less likely to be enrolled as cooperative members.

### Extension Visits

The most common services in Kenya are the home visits, conducted by extension officers, which allow for personal contact and specific recommendations. This costly administrative strategy is, however, rife with discrimination potential. Group extension demonstrations made at local farmers' plots offset tendencies for officers to skew visits to the privileged and are effective to the extent that farmers are aware of the meeting and able to attend. Yet few group demonstrations are given in western Kenya.

Findings indicate that female-managed farms always had less access to extension visits and demonstration plots than jointly managed farms (See Tables 1, 2, 7 & 8. This can be explained by the overwhelming predominance of male extension officers (in a culture where men and women strangers avoid contact), symbolic male authority over households despite extensive male absence, and the tendency for announcements about demonstration plots to be made at community meetings largely attended by men. Presumably, the presence of more staff should increase access to services for groups heretofore underserved, without depriving those who already have access. In the intensely served area, farmers received more visits, though disparities remained.

### Training

Courses at the nearby farmer training center provide a one to two-week intensive experience for farmers. It is rare for farmers to complete a course, but jointly managed households were four times more likely to have a member trained than female-managed farms. (See Tables 3, 9.) The minimal access for female managers is not surprising, given the tendency demonstrated in the research sample to pass information about opportunities through male communication networks and to use training as a way of conferring status. Moreover, female managers faced difficulties arranging for household responsibilities and cultivation tasks to be carried out in their absence, as they could not rely on resident spouses. In the intensely served area, farmers in general were only slightly more likely to receive training.

### Loans

Few of Kenya's small farmers received Guaranteed Minimal Return crop insurance loans, or loans from the Agricultural Finance Corporation. In the case of the former, minimum acreage requirements made virtually all farmers ineligible, as for the latter, a guarantee such as a land title deed or regular salary is necessary. Land reform was only completed in this part of Kenya in 1973, but since then it is mainly men -- in whose names title deeds are overwhelmingly held -- who have access to this valuable service. Applying for a loan is a long and complex process, intricately connected to local political contacts with authorities and institutions. While only 2 percent of those in jointly managed farms acquired loans, an additional 12 percent applied for or could correctly describe the complex application process, in contrast to the 1 percent of female managers who knew the application process. (See Tables 4, 10.) Given women's greater responsibilities for the production and storage of food for family substenance, their need for loans is acute.

### Special Service to Saturated Areas

The multipurpose cooperative society saturates services in the provision of easy access to maize seed, fertilizer, grade cow loans, and tractor services. Once enrolled in the cooperative, farmers experience less differentiated access to maize seed and fertilizer loans, as compared to the larger sample; female managers had slightly more access among the few who used the short-lived tractor service (Table 14, 15). However, for the valuable grade cows, a service tied closely to the cooperative political process, jointly managed farms with a man present were twice as likely to obtain a cow than female managers, even though equal numbers applied (Table 16). Overall, female managers were less likely to be recruited into the cooperative and thus to benefit from the service (Table 13).

### Supplying Information

Information about hybrid maize, the most important innovation in the area, spread differently to households, depending on whether a man was present. Members of jointly managed farms were more likely to hear about hybrid maize and its husbandry practices directly from technical sources, while women generally learned about them from their neighbors, members of women's groups, and traders (Tables 5, 11).

### Performance Under Different Settings Compared

What is the significance of these inequalities for farmers' performance? A controlled comparison of the minimally served Shikulu and the more intensely served, historically advantaged Shitoli area reveals that under conditions of unequal, male-oriented agricultural services, female managers experience a decline in their relative farm performance compared to men and a decline in their absolute performance compared to female managers in the ordinary area (Table 17). However, land scarcity also plays its part in this decline. With land being less scarce and with access to minimal services in the ordinary area, female managers are able to share labor with other women in their rich organizational networks, diversify, and adopt innovations faster than farms with a man present. In the intensely served area, the adoption rates of particularly appropriate crop mixes for female managers integrated into the extension delivery were double the rate on jointly managed farms (Table 18). When performance is not distorted by inequities, female managers' notable performance has also been documented in other studies. For example, research on yield per acre just south of this site found that female managers had higher productivity than men when access to education and extension were held constant.

## EXPLANATION AND IMPLICATIONS FOR POLICY IMPLEMENTATION

Extraordinary levels of services seem necessary to compensate for the gender differentiation that occurs under ordinary implementation, but this is achieved at the cost of farmers' dependency on costly administration. Three explanations for such differentials are set forth in this study, based on observation and research. First, prejudicial attitudes and ideological biases against women persist -- even against innovative, wealthier women managers farming on relatively large land sizes. Second, women are conceptually and politically invisible to the administration. As individuals they rarely initiate contact with the administration because of the suspicion and obligations they might incur, and in groups, they tend to operate autonomously of local politics, except for occasional linkages through wealthier women, whose priorities differ somewhat from those of low-income women. Third, administrators find it convenient to continue long-standing practices of neglecting women managers.

The lesson which emerges from these observations is the need to transform agricultural services if female-managed farms are to be reached, rather than simply to augment existing practices or to intensify services. Policymakers might consider three sets of alternative recommendations (see Box 1) designed primarily to affect the point of contact between farmers and field staff. Virtually all recommendations are cost-free, except for the energy needed to alter administrative inertia. The three strategies build on one another, leading from improved procedures to spread responsibility among men and women professionals, paraprofessionals, and community groups. Recommendations are broken down into categories of staff and staff training, extension and farmer training, credit, and cooperatives, and are summarized below.

In the first, improved management strategy, administrators might alter the incentive structure in which field staff work to reward effective and equitable administrative performance among the largely male staff who now tend to avoid women managers. Administrtors could further alter the composition of agricultural staff to recruit more females to serve as agricultural rather than home economics staff. Moreover, greater emphasis on group extension, to both men's and women's groups, would expand coverage in equitable ways at lower costs than individual home extension. Staff should also assure that information about training opportunities passes through women's networks as well as men's. Administrators might also establish minimum quotas to increase the percentage of loans to women managers, and examine application requirements to remove the discrimination that results from a land reform in which primarily men's names were placed on title deeds. Cooperatives should be encouraged to recruit women managers specifically and to call for joint membership when partners live together.

Using the second strategy, which builds on the recommendations above but also alters structures slightly, administrators might consider extending extension coverage with low-cost men and women paraprofessionals who live in areas they serve. Simultaneously, staff from Farmer Training Centers should be brought to the community to expand coverage to those who cannot easily leave farms for a week or two. At the same time, it must be recognized that home economics training is no substitute for agricultural training. Furthermore, alternative means to deliver credit could be made available in group loans to existing or newly formed women's groups, secured through group liability and consequent peer pressure. Lending facilities, mobile or located closer to farmers, would expand coverage to the less mobile, such as women managers.

The last strategy alters structures more significantly. Staff would work with community groups to share responsibility for decisions about agricultural change. Locally adapted research would be conducted according to a "farming-systems" approach. Agricultural and home economics extension staff and training would be integrated into a single service. And lastly, land reform would assure that producers' names, including women's are on title deeds.

Beyond these specific recommendations, other general policy changes should be considered such as fairer prices to farmers along with reduced or eliminated restrictions on trade across district boundaries, both of which would improve incentives for farmers. Women must also be integrated politically, as individuals and in groups, to learn about and receive the benefits or rural development.

# BACKGROUND TO THE STUDY

## AGRICULTURAL POLICY OBJECTIVES

Most Kenyans depend on agriculture and cattle production for their livelihood. Through policy, the government aims to supply information, inputs, and incentives to farmers to increase their productivity. To reach farmers, the agricultural ministry relies on a far-reaching agricultural extension service to deliver information and services to farmers. If extension fails to reach its clientele in terms which enchance productivity, then rural development does not proceed as effectively as it could.

The overall purpose of Kenyan agricultural policy is to improve yields of cash and food crops and thereby secure national self-sufficiency, more exchange earnings from export crops, and improved livelihoods for rural people. Kenya relies largely on the delivery of agricultural services to serve these purposes. The various policy objectives that flow from these purposes and services are listed below, including both national and locally specific concerns encompassed in the area of this case study.1

1. To assist all farmers to improve agriculture through a field-oriented extension service.

2. To provide specialized knowledge to farmers through formal training programs on an equitable basis.

3. To provide production loans to farmers on an equitable basis and, specifically, to encourage cooperative membership and receipt of cooperative loans.

4. To provide tractor services to Lirhembe cooperative members in an equitable manner.

5. To provide grade cows to cooperative members who apply for them.

6. To supply information to all farmers about hybrid maize and cash crops.

7. To decrease farmers' risk through adaptiveness and diversification.

In Kenya, as in much of Africa, women are the primary cultivators, managers, or co-managers of farm operations.

Women are responsible for later stages of land preparation; they also plant, weed, and harvest crops. Men, on the other hand, are responsible for early stages of land preparation such as clearing or burning bush, for plowing, and for cattle care. Yet despite women's extensive involvement in agriculture, Kenya shows the same tendency evident elsewhere in Africa: as has been noted in reference to Tanzania, "governmental agricultural policy has been designed as if women were not the majority of agricultural producers."[2] On the surface, however, official agricultural policy appears to be gender-neutral. Neither male preference, nor any other type of preference on the basis of region, economic status, or ethnicity, is officially encouraged.

This study examines the impact of agricultural policy on women in the Kakamega District in western Kenya in the 1970s. Before turning to look at the specific findings, it is useful to briefly review the process of agricultural implementation from the 1930s to the 1970s and thus obtain a historical perspective on women's agricultural roles and the origins of gender differentiation in agricultural policy.

## HISTORICAL PERSPECTIVES ON WOMEN'S AGRICULTURAL ROLES

The German anthropologist Gunter Wagner conducted an ethnography in the Kakamega District among the Maragoli and Bukusu subgroups of the Baluyia during the 1930s, providing a historical benchmark against which to assess change. Wagner related the descrpition of a Logoli elder (of the Maragoli group) on the sexual division of labor, and adds his comments:

> It is the wife's work to sweep, to grind, to cook, to build the fires, and to clean out the cattle partition. She carries water from the spring, buys the cooking-pots, and gathers firewood. She brings the salt (i.e., she burns the salt reeds in special pits and filters the ashes), cleans the walls of the house and the surface of the yard with cow dung, beats the floor of the house (so that it becomes hard and level), and 'knows' about the food on the food shelf.[3]

According to Wagner, "In addition to this list of household duties, the wife performs the greater and more strenous part of the garden labour, as hoeing is almost entirely woman's work." It is of interest that field work is missing from the elder's description, although his comment that women "know" about food may suggest some female role in managing the fruits of her labor. However, her obligation is to maximize the family's interest through sale or barter. Theoretically, a wife's profit belongs to her husband, but in practice, says Wagner, for a husband to inspect or question a woman's disposal of food was rare and considered meddling. A husband's main duties are, according to the elder,

> to care for his stool and beer pipe, to sweep his yard, to look after cattle, goats, and sheep, to know the place of digging, to cut the grass in the garden, to know when the roof of the house gets deteriorated and to pull the grass for thatching, to plait the string (for tying the grass) and to hand the grass (to the thatcher who is always a specialist).

Early government documents recognized women's predominance in agriculture, as did ethnographies, but reflected frustration in attempts to integrate men into farming. According to district agricultural files from the 1930s, "man has not the agricultural lore of the women."[4] Although token efforts to promote men's integration in agriculture with a sparse field staff were made, conscription, tax, and labor policy (the earliest of which involved forced labor) pulled what were termed "able-bodied men" from rural areas. Early preferential education for boys, which later provided men with the skills needed for occupational mobility, increased the likelihood that they would leave rural areas.

To stimulate production for the war effort during World War II, guaranteed high prices were offered for maize, resulting in substantial production increases and the emergence of the area as a major maize exporter.[5] With the absence of so many men during that time, it must be recognized that it was women who responded to price incentives.

Until the 1950s Emergency, which signalled an end to white settler farmer privileges, little stock was put in creating commercial farming among Africans, excluded even from growing certain cash crops such as coffee. The Swynnerton Plan, published in 1954, aimed to intensify agriculture through promoting cash cropping, mixed farming, and food crops, in that order of priority, through land reform, training, and extension. Secure individual titles, it was thought, would provide farmers with incentives to improve farms and with collateral for loans. In a sole paragraph, the plan called for women instructors at Farmer Training Centers (FTCs) to secure women's cooperation in agricultural development.[6] Land was consolidated and registered largely in men's favor; officials neglected women farmers' need for incentives to improve productivity, security, and obtain loans.

The plan also laid the foundation for a "progressive farmer" approach in which "able, rich, or energetic Africans would be able to acquire more land and bad or poor farmers less, creating a landed and landless class."[7] Such an approach meshes with the diffusion-of-innovation approach, which calls for attention to the risk-taking, open-minded minority that adopts innovations before others.[8] Both Swynnerton and diffusion-on-innovation perspectives ignore how political influence, wealth, and sex-separate communication networks distort this seemingly neutral process. Policy analysts affiliated with the University of Nairobi Institute for Development Studies later criticized the focus on the wealthy (some of whom were innovative, while some were not) with the least need.[9]

District archival documents reveal that agricultural officials were frustrated that they received few requests for agricultural assistance.[10] Given only a limited number of extension staff, who attended to those most likely to complain if unserved (the politically influential and/or wealthy), service distribution was probably quite skewed to the so-called progressive farmer. Yet during the 1950s, in conjunction with an expanded community development approach, group extension, including women's groups, reached its height in Kakamega, thus diluting the service delivery skew.

Descriptions of agricultural assistants visiting women's demonstration plots, women's groups, and women attending District Agricultural Officer *barazas* (local community groups) continued throughout documents of the 1950s. In Kakamega, documentation as early as 1951 described outreach to women's *barazas* using agricultural assistants as "propaganda agents" to spread the use of manure.[11] Both national and divisional documents reported employment of female agricultural staff. Reinforcing this, community development officers sponsored "homecraft institutes" which included the distribution of vegetable seeds, fruit trees, and talks on compost heaps. An extensive treatise on the acuteness of agricultural problems, attached to the 1955 District Annual Report, openly discussed men's alienation from agriculture and the need to teach women cash agriculture for their livelihood. Annual and division reports in 1959, when land consolidation efforts were very visible, noted that women's institutes "switched to practical matters, including homecrafts, making of dolls, land consolidation and enclosure, improved farm methods, and the sale of bean seeds."

The most significant change after independence (1963) was the development of hybrid maize at the Kitale Agricultural Research Station and its extensive promotion among farmers. Hybrid maize has the potential of raising output per acre from the approximately 6 bags (540 lbs.) achieved with local maize, to 35 bags, though in practice the best of farmers achieve not more than 20 bags per acre. Purchased seed, chemical fertilizer, and most importantly, labor-intensive husbandry practices, such as more thinning, weeding, and proper spacing, improved output. By the time of this research, officials estimated that 80 percent of Kakamega farmers used hybrid maize. However, despite the new demands on women's labor and management skills imposed by hybrid maize, women once again became invisible in agricultural policy documents, and the absence of agricultural outreach strategies to women is notable.

Beginning in the 1960s, passion fruit and tea were also promoted. "European" dairy cows, which double milk production with labor-intensive care, were introduced to farmers with loan assistance through the Agricultural Finance Corporation (AFC). Crops proved risky for farmers to adopt. For example, passion fruit seedlings and special loans for fertilizer and equipment were provided to farmers, along with a promise to open a juice factory in the Kakamega district once 500 acres were planted. By 1975 only half that acreage was planted, and fruit was rotting on the trees for lack of a market.

An extension-oriented approach continued after independence, but services were specialized in a manner that increasingly divided men and women into agriculture and home economics. With help from the U.S. Agency for International Development, a Home Economics Division was created in the Ministry of Agriculture, and training in the Unites States was provided for the top two tier officers along with a special three-mouth course at Egerton College for third-tier home economists. Normative conceptions about the desirable place of farm women are evident in the U.S. advisor's remarks:

> As land ownership and the desire to utilize land for cash cropping becomes more important in the African areas, it is believed men will assume an increasingly more responsible role in the actual day-to-day operation of the farm. Thus farm women will have an opportunity to spend more time with the care of their children and home. 12

It appears then that Kenya's agricultural extension approach in its initial and recent phases (at least up through the seventies) was founded upon some implicit assumptions about the extent and desirability of women's roles as farmers. A number of studies, including the present one, are finding these assumptions invalid. 13

## THE PURPOSE AND DESIGN OF THE STUDY

The purpose of this study is to compare the access of female-managed farms to agricultural services with that of jointly managed farms (which have a man present) to determine whether there are inequities in the delivery of agricultural services. The study also explores reasons for bias in service delivery, focusing on historical agricultural policy priorities, the interaction of class and gender, the administration of delivery, farmers' performance, and women's demands -- both individually and collectively -- for agricultural services. Finally, the study examines the consequences of inequitable delivery for women managers' agricultural performance.

### Research Methodology

As noted above, extension services are the major vehicle through which rural farmers are encouraged to adopt new crops and more productive farming methods. These services are offered at three levels of intensity: ordinary, intensified, or saturated. In ordinary implementation, agricultural field staff -- approximately one for every one or two sublocations -- provide extension advice to individuals in home visits and to groups in "demonstration plots." The agricultural instructor, assigned to the sublocation level (the smallest administrative unit), has the most frequent contact with farmers. He or she reports to a technical assistant at the location level, who in turn reports to a divisional officer. Though both men and women are employed, 98 percent of the field staff are men.[14] Agricultural instructors also provide information about credit and short courses at Farmer Training Centers.

Where ordinary services are intensified, line staff share extension tasks with specialists, increasing the likelihood that farmers will have contact with varied staff. The extension plan also called for one home economics officer per location, though in practice the ratio is often far less (for example, in Kakamega District only 4 of the 17 locations had a home economics officer). In saturated settings, not only is the ratio of staff to farm households higher, but greater institutionalized support for innovation and production is available. This includes services such as credit, training in efficient crop techniques, assistance in the introduction of new crops (with the provision of free maize seedlings, for example), labor-saving technologies and technologies to assist in the intensification of farming, and the establishment of cooperatives.

The current study, based on field research in western Kenya from 1974 to 1975, examines all three levels of implementation. The intervening years have seen some changes which are not reflected in this study. However, many of the basic policy, cultural, and environmental conditions described above here still prevail not only in western Kenya, but also in many parts of east and southern Africa. This study seeks to offer new empirical information and an analaysis that might usefully inform agricultural extension policy in general.

Survey data is drawn from 212 households in two sublocations of Idakho, Kakamega District: Shikulu and Shitoli (see Appendix and Note 24 for sample selection and data analysis procedures). After a brief description of the sublocations and of women's roles in the area, the study looks at services implemented on an "ordinary" basis in Shikulu to understand typical patterns of service delivery. Then access in Shikulu is compared with access in Shitoli, where services are intensified, to assess whether the intensification of staff and resources affect disparities. Finally, a subsample of Shitoli sublocation, consisting of members of the Lirhembe multipurpose cooperative society, is analyzed to determine the effects of saturating services on access.

Data on farm diversification also are compared for both sublocations to assess the impact of inequitable service delivery. This aspect of the study is based on a quasi-experimental design. Two administrative units, similar in culture, economy, and geography, but different in administrative intervention, allow qualified judgments to be made about programmatic causes of differential behavior.

Participant observation stems from residence in Kakamega District for seven months, and more specifically in Shikulu sublocation, where the Andala family graciously welcomed me as an additional member. Finally, documents in the Kenya government archives and secondary sources were analyzed over the course of four months in Nairobi, where I was an affiliate of the Institute for Development Studies at the University of Nairobi.

## Hypotheses

The specific hypotheses tested in this research are listed below:

1. The access of female-managed farms to agricultural services is less than that of jointly managed farms. (Subhypotheses to be tested involve specific services such as individual visits from extension officers, attendance at demonstration plots, training, and credit.)

2. The more valuable the service, the greater the gaps in service provision between the two types of farms.

3. Intensification of services improves female managers' access to services relative to female managers elsewhere, but gaps between them and jointly managed farms remain the same.

4. Female managers' farm behavior has no effect on the delivery of services.

5. Male preference in agricultural service delivery reduces women managers' agricultural performance.

6. Women's labor groups which share information and labor offset systematic biases in the delivery of services under non-landscarce conditions.

7. Female managed farms make fewer demands for agricultural services than jointly managed farms, both individually and collectively.

8. Women's exclusion from the political mainstream reduces government accountability to women and consequently lessens commitments to deliver services equitably to women.

9. Gender of manager surpasses economic status or land size as a cause for service delivery differentiation.

## THE LAND AND THE PEOPLE

### The District

The Baluyia people, a densely settled agricultural people who form the third largest ethnic group in Kenya, populate Kakamega District, western Kenya. They are comprised of relatively autonomous subgroups, each containing a loose collection of patrilineal clans, but they share both a common Bantu language, Luluyia (with some dialectical variations), and similar customs. The Idakho people, organized into 30 clans, are one such subgroup, located in the south central portion of the district. Idakho location contains nine sublocations, including Shikulu in the west and Shitoli in the east. The largest clan, the Bashimuli, resides in Shitoli; they have provided leadership for the location for nearly a half century, from Chief Sivachi to his son, Chief William.

With a population numbering three-quarters of a million, Kakamega is the second most densely settled district in Kenya, with 220 persons per square kilometer. 15 The density is partially responsible for the many men who migrate to large farms in the White Highlands and to urban areas in pursuit of wage employment. Male out-migration was 30 percent in 1937 and reportedly doubled by the 1950s. Since the post-World War II era, Kakamega district has been the largest labor-exporting area in Kenya, with the most uneven sex ratio. 16

The government labels Kakamega District a "high potential" agricultural area, given its annual rainfall of 70 inches per year, which permits two growing seasons. Maize and beans, the most common staple food crops, are grown along with millet, sorghum, root crops, and cowpeas. Virtually all farmers keep some chickens, used for feeding visitors. Only in recent decades have women started eating eggs, thought earlier to cause barrenness.

## Sublocations

Ten miles from one another, the two sublocations of Shikulu and Shitoli are similar in most agricultural respects. Both areas are populated by the Idakho subgroup of the Luyia people. Climate, rain, and soil features are similar, and the same proportion of men work away from home, leaving women to manage the farms. Both areas had the same mean economic standing, based on a five-point scale of material consumption items. 17 The average farm size was 2-1/2 acres, though it was smaller in Shitoli (1 1/2) and larger in Shikulu (3). Approximately half of the farm households owned a cow, but low cash levels and limited grazing land limit cattle holding. Although coffee has been a cash crop since Kenyan independence in 1963, it was grown by only 15 percent of the farmers in the study sample.

Finally, both sublocations were subject to similar mission influences, and schools were introduced in similar time periods. Approximately half the population in each sublocation identifies itself as Catholic, with the remainder being Protestant.

The key differences, therefore, relate to the amount of agricultural services available.

### Shikulu: Ordinary Services

Shikulu is the less well served area, receiving *ordinary* services. One agricultural instructor served approximately 2,000 farm households in a large territorial unit of 39 square kilometers. Recently transferred into that sublocation (transfer rates are high), he was nevertheless from Idakho location and thus spoke the local dialect. He had no bicycle, but relied on occasional vehicle traffic or the two buses which traveled each morning and afternoon to and from the tarmac road. Like the author, he spent much time traveling on foot. Shikulu is isolated from the administrative mainstream, being the furthest sublocation from the location center, division headquarters, and tarmac road from Kisumu. The nearest FTC, at Bukura, was located in the neighboring location, about ten miles away.

### Shitoli: Intensified and Saturated Services

In great contrast to Shikulu and to most sublocations in the district, Shitoli had the equivalent of three agricultural instructors, who served approximately 1,500 farm households. Its smaller territorial unit of 9 square kilometers is more densely settled than Shikulu (714 per square kilometer, compared with Shijulu's 225). Administrative offices are located on Shitoli's boundary; this has led historically to closer administrative contact between Shitoli and the government.

In addition, with assistance from the local Member of Parliament (M.P.), who is from the area, the Danish-supported Lirhembe multipurpose cooperative society was established in Shitoli. Cooperatives in Kenya as in other settings are established as an instrument for integrated small farmer development. They serve other secondary functions such as political mobilization repayment of political debt, and they also serve to establish a government presence in remote area. Whether these additional functions are viewed as appropriate or problematic, the aspect of primary importance to this study is the provision of a structure through which services can be delivered. For example, members had access to easy maize seed and fertilizer loans, to a tractor, and to high grade cows. (A second cooperative, for coffee, also existed in the Shitoli area, but was not examined in this study.)

Thus the cooperative structure in Lehrembe resulted in a *saturation* of services.

## A BRIEF DESCRIPTION OF WOMEN'S ROLES IN THE STUDY AREA

The data in this section are drawn from historical documents, other published research, and the current findings. It is intended to provide a brief view of how women operate in the research area.

### The Farm Unit

As a result of the changes in the economy outlined earlier, men now migrate as a normal part of their adult life cycle. Thus households, for the purpose of this study, have been divided into female- and jointly managed farms. The women in female managed farms were either widowed or separated from their husbands who had migrated outside the area for wage employment. Jointly managed farms consisted of husband-wife teams, or corporately managed farms with intergenerational management because land had not yet been parcelled out to sons. Forty percent of the 212 sample households consisted of female managers, a figure that corresponds very closely to the 36 percent for the district. [18]

Women with absent husbands faced several types of circumstances. Some had been all but deserted by their husbands, who permitted them to use the land and who retained patrilineal claims over children. Though financial assistance is difficult to calculate reliably, nearly one-third of these married women with absent husbands appeared to be getting no financial aid from husbands. [19] Still another group of women were widows who were self-sufficient, but were often assisted by children. These women had an added, nontangible benefit of the prestige that comes with age.

In the sample farms with a man present (60 percent), four distinct patterns were evident in nearly equal proportions. Some husbands were locally employed in government or owned a shop. These men may have aided in farm decision making, or transmitted information from male communication networks and contacts with political and agricultural influentials, but they did little actual farm work. Others had retired from wage employment, but they arbitrated local conflicts and did not do farm work. A third group consisted of men who had developed an interest in farming; some had even purchased plows. These men may have been genuinely interested in farming or simply reconciled to farm work, given the poor prospects for wage labor. A fourth group of men showed no interest in farming. Often on small farms, there simply is not enough work to keep two persons occupied, except during peak labor periods. On these farms, men appeared to be infected with a sense of malaise and discontent about chances for cash-related upward mobility. A significant number of small farms with a man present are among the poorest.

Many female managers absorbed men's responsibilities and coped with peak labor periods without the assistance of another resident adult. In the sample, half the female managers were responsible for cattle care. Also, many hired laborers and negotiated to have their land plowed prior to planting season. Four women were even managing a plowing business their husbands had begun. Men, by and large, "worked with money" if they could; that is, they sent money for oxen plowing and commercial seeds on which farmers now depend. Yet most husbands worked in low-paying jobs, as night watchmen and domestics for example, and their small salaries were quickly spent.

The patterns outlined suggest that female-managed and jointly managed farms stagger all levels of the local socioeconomic hierarchy, measured by income and land. Fifty-eight percent of women farm managers and 54 percent of farms with a man present had a low economic standing; similar proportions hold small acreage. Women farm managers inflate the low acreage category slightly because widows are part of the group.

## Decision Making

Women frequently make many of the agricultural decisions about husbandry and storage practices, crops grown, and time of planting; most of these agricultural activities are within the women's realm, even in farms with a man present. For example, in response to questions about when hybrid maize was grown or what practices are used, some husbands openly deferred to their wives. In a subsample of 107 farmers questioned about initial decision making connected with hybrid maize, 34 percent of the decisions were made by women, 31 percent by men, and 28 percent by both spouses (in 7 percent of the cases, a son made the decision).

Women without independent decision-making responsibilities were generally young and insecure. Contacts and influence in this society are solidified by blood and clan ties. Women, as in-marrying strangers, do not have those kinds of ties. After years of child-bearing and residence in her husband's clan, a woman is perceived to be somewhat linked to that clan, but for all practical purposes (and as demonstrated in the exogamy rules for her children's marriages) she is not a member of her husband's clan and never will be.

## Control Over Income and Income Responsibility

Women have the traditional, though informal, right to dispose of their produce. Male absence enhances this control over produce. Thus, women have some access to money, but in a context of ever-constricting land size, as land is divided among sons. That, combined with large families that must be fed, means that the wisest strategy often is *not* to sell the family's primary source of food. Women obtain money from trading, brewing beer, selling produce, and from their working husbands. Only 2 percent of the women in the sample earned a salary from government or wage employment, such as teaching. Fourteen percent of the women regularly brewed beer (*busaa*). Regular trading (buying and reselling) was an activity of only 6 percent of the women in the sample. Female traders were aware of the disparities in access to cash between them and male shopowners, who can acquire government loans. As one said, "Men have thousands (referring to shillings); women have hundreds." [20]

Women, thus, have considerably less access to and control over income than men. Almost all men have worked for wage employment at one time or another. Nor can it be assumed that family income is shared. Men rarely contribute more than occasional meat or milk for day-to-day household maintenance, and are not expected to do more. Household maintenance is women's financial responsibility. This includes providing for the basic needs of the family, including food, water, and clothing. Men are expected, however, to pay children's school fees and finance the plow operation.

## Access to Land and Credit

Patrilineal rules of inheritance still confine women's access to wealth. Land consolidation and registration, begun in the late colonial era, were completed for most of Idakho in the early 1970s. Although in earlier eras land had been managed and passed patrilineally, the traditional process had nonetheless guaranteed secure access to women cultivators. Land titles, now held largely in men's names, are alienable; a full fifth of the sample claim land changes. Women now work in a tenant-like relationship to land, as no concept of joint tenancy exists. Elsewhere in Nyanza Province among the Luo, a study found women owning only 6 percent of land parcels after registration. [21]

With no control over land, women lack collateral for loans. Therefore, if a woman wants a loan, her husband must put up a guarantee in the form of a title deed or salary, which can be difficult to arrange if husbands are not regularly at home, or are not interested in the farm. The only genuinely independent women for loan purposes are widows, who can arrange legally, through a complicated and costly process, to have the title deed transferred to their names. Putting title deeds in male names has solidified male control over powerful resources, and this has obvious implications for acquiring other resources as well.

A man can obtain a loan without his wife's awareness, yet loan default and subsequent land auction mean women have no access to land to produce for their children or for a livelihood. In the area, an elite woman widow was threatened with foreclosure when she became liable for an agricultural loan her late husband acquired (without her knowledge) and squandered for nonagricultural purposes. This kind of incident occurs with enough frequency to be cited by women as a major source of anxiety. 22

## Organizational Activity, Political Authority, and Access to Agricultural Services

Organizational activity among women has flourished. Over 90 percent of the women in the sample belong to one or more groups. All these groups, complete with names and officers, are recognized by local residents. Women continue the tradition of collective labor in agriculture, in *buhasio* groups, where members work for a fee, divided at year's end, and for food and tea with lots of sugar. Women also belong to *bulala* groups for mutual aid in times of crisis; the aid is collected in kind or in cash. Women, who dominate church membership in rural areas, form groups for church fund raising or for praying. While men are part of other networks, sit on government committees, and hear announcements, few are sustained members of groups like women.

On the whole, though, women's organizational activity is not integrated into the political mainstream. During elections, women's groups or their leaders support candidates with promises of votes in exchange for short-term benefits, such as matches or packets of sugar. But women are neither strategically located for influencing local administrators nor included in information dissemination networks. Formal political authority and communication networks emanating therefrom are largely men's affairs. Women are rarely in positions of authority, except to represent women. Furthermore, in communities like these, which have undergone substantial economic differentiation, a woman in formal authority is usually of elite status and often has different interests, perceptions, and priorities from other women.

Because of agriculture's importance in the local economy, agricultural services, credit, and staff contacts are central to power in Idakho. The agricultural instructors and technical assistants frequently meet and advise local administrators as well as meet weekly with the subchief and chief at *barazas* (local meetings). Thus, the agricultural field staff belong to a communication network that dispenses information and material resources, subjecting them to political contact and pressure. In these networks, information about loans and training is passed, and valuable contacts are made.

# FINDINGS

## IMPLICATIONS FOR KENYA AGRICULTURAL POLICY IMPLEMENTATION

This study, like others in the series, presents findings in relation to policy objectives. The following section analyzes the implementation of Kenya's agricultural policy objectives as outlined in Part I, but presented here in a different order. Objectives relevant to ordinary implementation in Shikulu sublocation will be examined first. The effects of intensification and saturation in Shitoli sublocation will be examined next. To obtain a broader picture of the effects of saturation, the impact of objectives which were relevant only to a saturated setting will be considered as well. Finally, a comparative analysis of women's performance in all three settings will be undertaken, and considered in relation to the broad objective of increasing adaptiveness and diversification.

## ORDINARY IMPLEMENTATION; SHIKULU SUBLOCATION:

### IMPROVING AGRICULTURE THROUGH A FIELD-ORIENTED EXTENSION SERVICE

As indicated earlier, in Kenya extension agents deliver and adapt information about agricultural research to farmers in their homes and at demonstrations on local farmers' plots. Also knowledgeable about training and credit, the agent can link these beneficial services to farmers. The government aims to have one agent per sublocation, although in practice one may serve two or three sublocations. Extension agents focus on hybrid maize and its associated husbandry practices, and on cash crops, such as sugar cane, coffee, European vegetables, and passion fruit, and encourage crop diversification so that farmers spread their risks, should one or several crops fail.

Home extension visits allow individual, personal, and sustained contact on the farm itself, where recommendations can be made to suit farm conditions. The disadvantage of home visits is that they are costly and rife with discrimination potential. It is also difficult to monitor the quantity of individual home visits, as field officers operate with great freedom. Studies in Kenya have found that extension officers visit an average of only 4-5 households per week. 23 These studies also show that a quarter to half of the extension officers were misinformed on crops about which they extend information, and that officers work in an administrative structure with few incentives for improving or making their performance equitable. Some farmers in the current sample spoke disparagingly of the minimal advice agents had to offer or of contradictory instructions they received.

Group extension, through lectures at demonstration plots on local farmers' plots (including those of some elite women farm managers), offsets the tendency to skew visits to few farmers by opening up services to all. Demonstration plots can be educational in practical and experiential ways, because farmers visualize new tasks, such as how to apply fertilizer, and can then later view progress on the local plot. Agricultural instructors are more likely to prepare for the presentation they make, and such events are easily monitored (perhaps accounting for the better preparation). Yet demonstration plots in Western Kenya are few and far between, perhaps once a growing season in a sublocation, partly due to the preparation time association with acquiring a plot to use, finding inputs, and publicizing the event through announcements in *barazas*. Attendance is based on the initiative of the farmer. In the past farmers were often compelled to attend.

Historically, compulsion was used in other ways as well: seedlings not planted in lines would be uprooted, farmers would be fined for improper coffee care, or could face compulsory labor. As part of efforts to eradicate soil erosion, farmers were often forced to work on bench terracing. An ongoing resentment and suspicion of agents was sometimes reflected in farmers' remarks.

## Home Visits

Farmers were asked whether their farms had ever been visited by an agricultural instructor. Their responses, presented in Table 1, show that female-managed farms were significantly less well served than jointly managed farms. About half the female-managed farms had never been visited, in contrast to only a quarter of the jointly managed farms.

Table 1. Visits by Agricultural Instructor, by Farm Management Type [24]

| | Female-Managed | Jointly Managed |
|---|---|---|
| Farm never visited | 49%/42 | 28%/36 |
| Farm visited once or more | 51%/43 | 72%/91 |
| (Yules Q=.42; Significance=.01) | | N=212 |

How did women fare in jointly managed farms? Because many husbands were not involved in farm work and not at home during the day, often women were in fact alone at the farm when agricultural instructors visited, and agricultural instructors generally spoke to "whomever was there." When the husband was present during a visit communication patterns were diverse, depending, as one might expect, on the personality styles of the couple and the agricultural instructor. While the general norm was for the husband to speak and to represent the household, this did not necessarily mean that the woman left or stayed silent; thus women in jointly managed farms may have more frequent and direct contact with agricultural instructors than women managing farms alone.

It is difficult to determine, however, whether or not the quality, intensity, and duration of the communication between an instructor and lone woman, one whose husband is away for the day, compared favorably with that between an instructor and husband. This probably also varied with personal style. It is also important to bear in mind that many middle-aged and older women have a strong sense of personal efficacy and considerable prestige derived from their reputation as hardworking farmers and mothers of many children. Thus, with increasing age, the potentially problematic nature of communication between women and instructors becomes less important.

### Demonstration Plots

As Table 2 indicates, fewer farmers have attended plots than have received visits, but the disparities between management types were less marked than in visits.

Table 2. Demonstration Plot Attendance by Farm Management Type

| | Female-managed | Jointly managed |
|---|---|---|
| No. household member attended | 62%/52 | 46%/59 |
| 1+ household members attended | 38%/32 | 54%/68 |
| (Yules Q=.31; Significance=.05) | | N=211 |

Lower attendance rates for woman managers could mean that they were uninterested in, unaware of, or unable to utilize this service. In this context, it becomes relevant that demonstrations were usually announced at the weekly barazas, posing a problem for women who rarely attend those because of time constraints and custom (it has traditionally been the man's responsibility to attend). Barazas last several hours as announcements are made and local disputes adjudicated. The percentage of women who can spare such time rarely exceeds 10 percent.

Among jointly managed farms, equal numbers of men and women attended demonstrations, the explanation for which is twofold. First, male presence brings a farm directly into a valued male communication network. Second, the presence of two or more adult laborers on a farm permits the release of one or the other from farm and child-care activities to attend the meeting.

## PROVIDING SPECIALIZED KNOWLEDGE TO FARMERS THROUGH FORMAL TRAINING PROGRAMS ON AN EQUITABLE BASIS

Another service, training, was provided at farmer training centers, which hold short courses that usually last one to two weeks. Though highly subsidized by the government, a fee of shs. 10/50 (approximately $1.50) is required, and this is a sizeable sum for farmers without a regular cash income. [25] Training represents a direct and intensive service for farmers, as it lasts longer than other services and makes available highly qualified teachers. National and District Reports since independence indicate that a third of all FTC participants are women, but they are virtually always tracked into home economics courses covering a wide scope of topics, with at best a third of course attention relevant to agriculture.

Bukura Farmer Training Center is a 5-15 mile trip from farms in both sub-locations sampled, and has been open since 1923. As the results in Table 3 reveal, completing a course at the farmer training center is rare, a surprising finding given Bukura's proximity. But a full third of the farmers were unaware that the training center existed. Aside from that, great disparities exist, with jointly managed households being four times more likely to have a member trained than female-managed households.

Table 3. Training by Farm Management Type

| | Female-managed | Jointly managed |
|---|---|---|
| None in household trained | 95%/80 | 80%/102 |
| Someone in household trained | 5%/4 | 20%/25 |
| (Yules Q=.66; Significance=.01) | | N=211 |

For women managing farms alone, a one- to two-week training period presents special problems. They must arrange for day-to-day household responsibilities and cultivation to be provided for while they are away. In the 84 female managed farms in the sample, only four women had ever had any training.

Chiefs and assistant chiefs help agricultural instructors recruit and select trainees through announcements at barazas or through word of mouth. Many farmers perceive that attendance at Bukura is an honor, achieved by invitation only. The awareness of and opportunity to attend training is thus more closely linked to the local power structure than the first two services considered.

In addition, jointly managed farms, husbands were often wary about wives being away for extended periods of time, and, in some cases, the chief or assistant chief had to persuade husbands to allow their wives to attend. Of the 25 jointly managed farms where some member had attended a training course, six of those trained were women, and the remainder were husbands or sons.

But it is noteworthy that in a number of cases where sons were trained, they had since moved away from the farm and found non-agricultural employment. In several cases where husbands were trained, the men were old and, for all practical purposes, retired from active farms work; their invitation to be trained simply conferred status.

## PROVIDING PRODUCTION LOANS TO FARMERS ON AN EQUITABLE BASIS

The earliest type of loan available to respondents was the Guaranteed Minimum Return (GMR) advance for maize seed and fertilizer, but the minimum acreage required for maize was beyond the holdings of most ordinary farmers in the location. Agricultural Finance Corporation (AFC) loans became available after land reform was completed in 1973, when farmers could acquire individual land title deeds. Most lending benefits the large farm sector. Kenya's official plan indicates that 1 percent of smallholders receive these types of loans. 26

Loan application and acquisition procedures integrate farmers more fully into the extension service. In addition, the district-level personnel involved in administering loans were of higher caliber than the ordinary agricultural instructors.

Table 4 shows whether or not respondents knew about loans and loan procedures, or had actually received a loan, by farm management type. In light of the above information, it is not surprising that ordinary farmers acquired only three loans, and these went to jointly managed farms. An additional 7 percent (16 households) could correctly relate the application process. Only one of the 19 households that knew anything about loans was female-managed, and that particular woman was not only wealthy, but was linked to the local power structure by participation on the development committee, marriage to another committee participant (though he was absent), and an in-law relationship to the local politician.

Table 4. Loan Knowledge and Acquisition by Farm Management Type

| | Female-managed | Jointly managed |
|---|---|---|
| Knew nothing about loans | 99%/83 | 86%/109 |
| Knew application process or had applied for GMR/AFC loans | 1%/1 | 12%/15 |
| Acquired either GMR or AFC loan | -- | 2%/3 |
| (Q=.86; Significance=.01) | | N=211 |

## SUPPLYING INFORMATION ABOUT HYBRID MAIZE AND CASH CROPS

Because of the special husbandry practices required to achieve high output from hybrid maize, the widespread acceptance of hybrid maize implies an immense communication task for agricultural instructors. These practices are not something farmers would adopt on the basis of experience alone: they involve detailed expertise on spacing, thinning, planting time, weeding size, and techniques for applying fertilizer.

When farmers who grew hybrid maize were questioned about their initial source of information about the crop, the results showed that communication channels for innovation are gender-specific. Almost half of the respondents from jointly managed farms heard about hybrid maize directly from technical sources, which was true for only a quarter of female-managed farms. Female farm managers heard about maize from neighbors in both formal and informal communication networks, the latter consisting mainly of communal agricultural labor groups.

Table 5. Sources of Initial Information about Hybrid Maize by Farm Management Type (Hybrid Maize Growers Only)

| | Female-managed | Jointly managed |
|---|---|---|
| Agricultural staff | 25%/21 | 44%/51 |
| Neighbors | 43%/36 | 33%/38 |
| Market/stockists | 25%/21 | 9%/10 |
| Other (employed outside; radio) | 7%/6 | 15%/17 |
| (Gamma= -.23; significance = .001) | | N=200 |

Women managers also heard about new crops from the market area, which they visit several times a week to sell produce, to grind maize for cooking, or to purchase household goods, and in particular from "stockists," the shop-keepers who sell agricultural inputs such as seed and fertilizer. Also at the market are further extensions of informal women's networks and groups, such as those that exist among women traders. (Traders often buy packets of seed and sell them in small portions to buyers lacking enough cash for a full bag of seed.) Such networks are useful in the diffusion of agricultural innovations, but shopkeepers located far in the interior purchase innovative inputs well after their initial introduction in an area.

Hybrid maize farmers were also asked about the husbandry practices they used and, when they responded with some proper technical practices (as did most farmers), the source of information was requested.

In this case, too, female managers received considerably less information from agricultural staff than did men. Less than half of the female managers had direct access to technical information from those trained in agriculture, in contrast to the almost three-fourths of persons from jointly managed farms with such access. Most female managers found out about hybrid maize practices from second-hand sources.

Table 6. Sources of Information About Hybrid Maize Husbandry Practices by Farm Management Type (Hybrid Maize Growers Only)

| | Female-managed | Jointly managed |
|---|---|---|
| Agricultural Personnel | 43%/35 | 72%/83 |
| Non-Agricultural Personnel | 57%/46 | 28%/32 |
| (Yules Q= -.55; Significance = .0001) | | N=196 |

## INTENSIFIED IMPLEMENTATION: SHITOLI SUBLOCATION

What are the effects of intensified services on equitable administrative performance? While increasing resources is not a sufficient condition to guarantee more effecive services, it has been argued that in theory, at least, administrative performance could improve in both quality and quantity with increased resources for staff and support services. Many problems in administrative performance are said to result from financial constraints under which development administrators operate. Low budgets, staff shortages, insufficient transport vehicles, and the like are given as explanations for the failure of administrators to service clientele adequately.[27] Moreover, bureaucratic innovation is predicted with increased resources such as time, finance, material, or skills.[28] Increased resources can supposedly enhance innovation by expanding an administrative clientele (to include, for example, women) without depriving those clientele already served.

The section below compares access in Shikulu and Shitoli in relation to the four objectives outlined above and illustrates some of the limitations of this approach.

## IMPROVING AGRICULTURE THROUGH A FIELD-ORIENTED EXTENSION SERVICE

As Table 7 indicates, both female-managed farms and jointly managed farms in Shitoli generally received more extension visits than all farms in Shikulu. However, disparities still remained between farm management types in the intensified Shitoli area, although the gaps were narrowed somewhat. Furthermore, it is noteworthy that intensification seems to merely raise the access level of female managed farms to that of jointly managed farms in typically served areas.

Table 7. The Impact of Intensification on Extension Visits by Farm Management Type

| | SHIKULU | | SHITOLI | |
|---|---|---|---|---|
| | Female-managed | Jointly managed | Female-managed | Jointly managed |
| Farm never visited | 61%/28 | 37%/23 | 36%/14 | 20%/13 |
| Farm visited 1+ times | 39%/18 | 63%/39 | 64%/25 | 80%/52 |
| | N=108 Q=/45 | | N=104 Q=.38 | |
| | Tau B=.24 (significant at the .05 level) | | Tau B=.18 | |

In the case of access to demonstration plots available data show that intensification of services did not substantially affect the gap between the farm types, as Table 8 demonstrates.

Table 8. The Impact of Intensification on Access to Demonstration Plots

| | SHIKULU | | SHITOLI | |
|---|---|---|---|---|
| | Female-managed | Jointly managed | Female-managed | Jointly managed |
| No household member attended | 72%/33 | 56%/35 | 50%/19 | 37%/24 |
| 1+ household member attended | 28%/13 | 44%/27 | 50%/19 | 63%/41 |
| | Q=.32 Tau B=.16 | | Q=.26 Tau B=.13 | |

## PROVIDING FORMAL TRAINING PROGRAMS

The existence of more staff in Shitoli should increase the likelihood that farmers hear about training. However, as Table 9 shows, overall only a slightly larger number of farmers (both in female-managed and jointly managed farms) were exposed to training with intensified services. Gaps between the sexes narrowed slightly, from 16 to 14 percentage points. But here the Shitoli female managers did not do much better than the Shikulu joint managers, suggesting the importance of communications and women's mobility constraints for this service.

Table 9. The Impact of Intensification on Training

| | SHIKULU | | SHITOLI | |
|---|---|---|---|---|
| | Female-managed | Jointly managed | Female-managed | Jointly managed |
| None in household trained | 98%/45 | 82%/51 | 92%/35 | 78%/51 |
| Someone in household trained | 2%/1 | 18%/11 | 8%/3 | 22%/14 |
| | Q=.81 Tau B=.24 (Significant at the .05 level) | | Q=.52 Tau B=.18 | |

## PROVIDING PRODUCTION LOANS

The convenience of more and nearby specialist staff in Shitoli should increase farmers' access to loans under the conditions of intensified services, but population density and consequent small farm sizes in Shitoli offset these prospects. The minimum floor of GMR crop insurance loans is far too high for even relatively large landowners in Shitoli to qualify, and AFC loans were commonly promoted for "grade cows" which require grazing areas of at least an acre planted for fodder grass. Thus, as Table 10 indicates, neither region seems to have been very informed about loans, and increased services had less of an impact than might be expected.

Table 10. The Impact of Intensification on Knowledge About and Acquisition of Loans

| | SHIKULU | | SHITOLI | |
|---|---|---|---|---|
| | Female-managed | Jointly managed | Female-managed | Jointly managed |
| Knew nothing about loans | 100%/46 | 81%/50 | 97%/37 | 91%/59 |
| Knew loan application process or had applied for GMR/AFC loans | -- | 16%/10 | 3% | 7%/5 |
| Acquired loan | -- | 3%/2 | -- | 2%/1 |
| | Q=1.0<br>Tau B=.30<br>(Significant at the .05 level) | | Q=.58<br>Tau B=.13 | |

## SUPPLYING INFORMATION ABOUT HYBRID MAIZE AND CASH CROPS

As Table 11 shows, in both sublocations same-sex communication is still a common vehicle for information transmission. Unexpectedly, a larger proportion of Shitoli jointly managed farms, a proportion nearly comparable to that among female managers, cited neighbors as an /initial/ source of information.

Table 11. Initial Source of Information about Hybrid Maize by Sublocation and Farm Management Type

| | SHIKULU* | | SHITOLI** | |
|---|---|---|---|---|
| | Female-managed | Jointly managed | Female-managed | Jointly managed |
| Agricultural staff | 20%/9 | 43%/25 | 31%/12 | 45%/26 |
| Neighbors | 42%/19 | 24%/14 | 44%/17 | 41%/24 |
| Stockists | 29%/13 | 14%/8 | 21%/8 | 3%/2 |
| Other | 9%/4 | 19%/11 | 5%/2 | 10%/6 |

*Significant at the .01 level.
**Significant at the .10 level. N=103 N=97

As shown in Table 12, the source of information for hybrid maize practices however, continues to vary according to management type. Though far more women in highly serviced sublocations learn practices from agricultural sources, the proportion was just close to the proportion of jointly managed farm representatives in an ordinarily served sublocation. Even in Shitoli, many women heard about practices from their neighbors, members of women's groups, and traders.

Table 12. Source of Information about Hybrid Maize Practices by Sublocation and Farm Management Type

| | SHIKULU | | SHITOLI | |
|---|---|---|---|---|
| | Female-managed | Jointly managed | Female-managed | Jointly managed |
| Agricultural staff | 30%/13 | 61%/35 | 58%/22 | 83%/48 |
| Non-agricultural | 70%/30 | 39%/22 | 42%/16 | 17%/10 |
| | N=100 | | N=96 | |
| | Yules Q=-.57 | | -.55 | |
| | Tau B=-.31 | | -.27 | |
| | Significance=.01 | | .01 | |

## SPECIAL SERVICES TO SATURATED AREAS: LIRHEMBE COOPERATIVE (SHITOLI)

At this point, data from the Lirhembe Cooperative, an even more intensely served subsample, are analyzed to determine whether the saturation of services offsets usual biases against women managers and addresses their special difficulties in densely settled Shitoli with its food shortages, dependency on cash, and burdensome work setting. High land density in this area makes it difficult for women to grow the quantity of food necessary for feeding their families without cash support.

## ENCOURAGING COOPERATIVE MEMBERSHIP

Lirhembe's provision of easy loans, grade cows, and tractor services to its members should address farmers' needs. First, however, women managers must enroll as cooperative members or convince generally absent husbands to do so. Thus, before examining female manager's access to loans, it is necessary to explore their propensity to join the cooperative.

Though female membership in the cooperative is not prohibited, managers assumed that male "heads" would make such decisions and sign for the family, and no recruitment effort was directed at women managing farms alone.

Table 13. Lirhembe Cooperative Membership by Farm Management Type

| | Female-managed | Jointly managed |
|---|---|---|
| Member | 46%/18 | 72%/47 |
| Non-member | 54%/21 | 28%/18 |
| (Yules Q= -.51; Significance=.01) | | N= 104 |

The figures for cooperative membership, displayed in Table 13, are striking. While almost three-fourths of the jointly managed farms in the Shitoli sample were registered at the society, less than half of female-managed farms were registered. In addition, in the subsample of 65 cooperative members, only 28 percent (or 18 farm households) are female managed. All the known females registered who comprise only 7 percent of members, were widows.

## ENCOURAGING COOPERATIVE LOANS

Cooperative maize seed and fertilizer loans were easy to acquire, compared to the government AFC loans. All one needed to do was join the cooperative, sign for the agricultural inputs, and agree to repay either in maize or in cash, presumably at the end of the growing season. No title deeds were required. In fact, people avoided payment at the end of various seasons, and declining numbers had repaid each year since the loan was offered. In 1975, no capital was available to purchase inputs to make loans available. The populace was massively indebted, but the indebtedness lacked any penalties. Terms of repayment were widely understood at the time of interviews (the fourth year of the program), but members perceived that the society was "not strict" about repayment. Little or no selectivity was involved in the provision of this loan.

Table 14. Use of Maize Seed and Fertilizer Loans in Lirhembe Cooperative by Farm Management Type

| | Female-managed | Jointly managed |
|---|---|---|
| Used loan | 78%/14 | 87%/41 |
| Did not use loan | 22%/4 | 13%/6 |
| (Yules Q= -.32) | | N=65 |

As Table 14 shows, jointly managed farms were still more likely to use loans than female-managed farms. Perhaps the long-standing custom for husbands to represent households and negotiate transactions outside the home explains the difference. Nonetheless, it was possible for women to sign for the agricultural inputs, even when membership was in their husband's name. In examining Society ledgers, approximately 20 percent of the signatures for inputs were women's names, while the number of women members, as indicated above, was only 7 percent of the total. Still, fewer women's signatures were found than expected, perhaps partly because of the long-standing custom among women, by choice or otherwise, of operating autonomously from government assistance.

## PROVIDING TRACTOR SERVICES TO LIRHEMBE COOPERATIVE MEMBERS

Another concrete benefit available to farmers was a tractor. Tractors could alleviate burdensome hand digging of the soil for cultivation, which is always women's work. Given widespread perceptions that this service was limited to persons with political influence--since committees rather than the co-op clerk made decisions--people were pessimistic about its accessibility. Table 15 shows the distribution of tractor services by farm-management type.

Table 15. Cooperative Tractor use by Farm Management Type

| | Female-managed | Jointly managed |
|---|---|---|
| Tractor used | 29%/5 | 23%/11 |
| Tractor not used | 71%/12 | 77%/36 |
| | | N=64 |

Somewhat surprisingly, similar proportions of female-managed farms and jointly managed farms received tractor services. Yet all in all, only a quarter or so of the farms in the subsample were serviced. Contrary to findings on maize and fertilizer usage, data on tractor service indicated that female-managed farms. In fact, the slightly higher proportion could be accounted for by this service being particularly in tune with women's needs. The tractor was available for two years, until its mechanical failure, at which time no capital was available for repair.

## PROVIDING HIGH-GRADE COWS TO COOPERATIVE MEMBERS WHO APPLY FOR THEM

One of the most valuable services offered to farms has been high-grade cows, which produce twice as much milk as local cows. At Lirhembe, grade cows were "borrowed" and repaid with young heifers which were then borrowed by other cooperative members on the same basis. Plans had been made to acquire a milk cooler for the central collection and marketing of milk. Milk sales were also to be used to repay the cost of barbed wire fences loaned to farmers who received grade cows, a requirement for this service. Funds for this cooler were not available, however, in 1975.

Program impact had been genuinely redistributive. Many poor farmers obtained a cow; these farmers would never have accumulated the cash to purchase such a costly animal. 29 And as the previous section demonstrated, for loan purposes the regular extension service had virtually written off farmers who did have sufficient, albeit small-sized, land.

Yet several requirements of the application process have skewed receipt in favor of farmers with more land. To receive a cow, farmers had to make available 1 1/4 acres for grazing and planting fodder grass. Besides that, farms had to have at least one acre for maize, in addition to space for their house, meaning that only farmers with 2 1/2 acres or more could obtain the grade cow loan, a size well above the 1 1/2 average acreage.

Furthermore, given the high value of a grade cow and the committee selection process used to determine eligibility for the cow loan, the potential for discrimination was high, be it political, economic, or gender-based. Not enough cows were available for the entire membership. Influence and contacts were thus essential for pressing or "reminding" committee members of an application. This is generally considered men's activity, particularly because it is largely men who sit on the committee.

Table 16. Cooperative Grade Cow Applications by Farm Management Type

| | Female-managed | Jointly managed |
|---|---|---|
| Applied; received cow | 22%/4 | 47%/21 |
| Applied; did not receive cow | 39%/7 | 17%/8 |
| Did not apply | 39%/7 | 38%/18 |
| | | N= 65 |

As Table 16 indicates, although three-fifths of both female-managed farms and jointly managed farms applied for cows, a manager of the joint type was twice as likely as a female manager to actually receive a cow. Female managers were not disproportionately concentrated on smaller land parcels, thus eliminating economic factors as the cause of inequity.

## PERFORMANCE UNDER ORDINARY, INTENSIFIED, AND SATURATED SERVICES COMPARED

The documentation of thoroughgoing inequities in the delivery of agricultural services is made more meaningful when linked to farmers' performance. If discrimination had no impact, then reforming or transforming agricultural administration would be beside the point. This section reveals not only that discrimination does have detrimental impacts, but that relevant yet untapped women's groups are vehicles for encouraging information dissemination, risk-taking, and labor sharing.

Policy impact analysis attempts to determine whether policies produce intended (or unintended) effects on a populace, but it is difficult to sort out policy from other effects. Ideally, impact studies are longitudinal, measuring baseline characteristics of a sample prior to the intrustion of a policy change and then after a designated time period, using the same instrument to measure change. In the absence of longitudinal data, controlled comparisons between two virtually identical administrative units -- one affected by the policy and the other not--permit qualified generalizations about policy impact.

Using a controlled comparison of Shikulu and Shitoli, this section tests the hypothesis that the performance of female managers relative to joint managers is declining under conditions of intensified, male-oriented agricultural services. The similarity of both sublocations, alike in all ways except for the intensification of services and land scarcity, compensates for the lack of longitudinal data.

To further compensate for the non-longitudinal design of the study, farmers were asked factually oriented questions about their past and present agricultural practices. Unless farmers sell maize in larger quantities all at once, they have little sense of output per acre in unit measures; much food, in fact, is stored after harvest in granaries. Therefore, performance measures used included crop diversification and timely adaptiveness to hybrid maize innovations. The more diversified the crops, the greater the spread of farmers' risks should one crop fail. The earlier hybrid maize was adopted, the greater farmers' adaptiveness.

The table below reveals high performance scores for Shikulu female managers -- indeed higher than that of jointly managed farms -- and low scores for their Shitoli counterparts. Female managed farms in Shitoli appear to be experiencing a relative decline compared to jointly managed farms, but also an absolute decline compared to female-managed farms in Shikulu. This reverses the pattern discovered in relation to other objectives; Shitoli female managers experience no improvement compared to their female counterparts in Shikulu except in diversifying cash crops, and their gaps with jointly managed farms are quite marked.

**Table 17. Mean Number of Cash Crop and Food Cro Types and Years Since Hybrid Maize Adopted by Sublocation and Farm Management Type** [30]

| | SHIKULU | | SHITOLI | |
|---|---|---|---|---|
| | Female-managed | Jointly managed | Female-managed | Jointly managed |
| Mean No. cash crops | 1.4 | 1.4 | 1.7 | 1.8 |
| Mean No. food crops | 2.2 | 2.0 | 1.6 | 2.0 |
| Years since hybrid maize adopted | 3.6 | 3.5 | 3.3 | 4.2 |

In Shikulu, with relatively few services both now and historically, female-managed farms either surpass or equal jointly managed farms in their crop diversification and in the earliness with which they adopted hybrid maize. In Shitoli, on the other hand, women score lower on all measures, and particularly on food crops grown for family consumption. This is noteworthy because food crops are customarily thought to be within the work domain of women. Jointly managed farms grow more diverse food crops, which may be partially due to their greater access to services, particularly the more recent ones which stress food.

Over the short term, women's commitment to and experience in farming results in rates of new crop adoption similar to those observed in jointly managed farms, which have better access to extension services. Indeed, the mean may actually underestimate female managers' abilities, because female managers operate under greater labor constraints than those faced by jointly managed farms. In addition, this comparative performance of Shikulu women is viewed against the low level of (male-oriented) service which appears to have minimal impact even on farms with a man present. In Shitoli, the male-oriented but generally intensified services over a period of several decades have incorporated both female and jointly managed farms into an agricultural administrative framework, making improved performance more dependent on service delivery.

Administrative services and land scarcity also contributed to explaining differences in farm performance under different levels of services. In Shikulu, an overwhelming majority of women participated in communal groups which share agricultural information and labor. Women's organizational activities enhanced female managers' performance. Furthermore, additional land enabled women to put ideas diffused through this network into practice, without risk to the basic food supply. Finally, there were also market outlets for income-earning activities. In Shitoli, extra land was simply not available and access to assistance from women's labor groups was also less available. Still, food crop diversification was maintained in jointly managed farms through continued administrative encouragement, which suggests that land scarcity does not spell doom for food diversification.

The second more important explanation for gaps is the systematic discrimination against women managers in a service delivery system characterized by male preference. The innovativeness of Shitoli jointly managed farms suggests that governments do play some role in farmers' performance which can offset the effects of land scarcity. Without the support of the agricultural staff, Shikulu women managers innovate independently by calling on group support in ways that men cannot. Once a farming population becomes more dependent on services, however skewed, the situation changes. After a period of systematic neglect in support services or valuable aid such as loans, performance deteriorates.

But how does one sort out the administrative versus land scarcity explanations with any confidence? If land scarcity were more important, one might very well "write off" Shitoli, yet the performance of jointly managed farms suggests such a judgment would be premature. It would also be wrong to under-estimate women managers under land scarcity conditions simply because their organizational resources and independent innovativeness are no longer as functional as they were. Indeed, with sufficient administrative support, women once again surpass men, as shown in Table 18.

Examining the saturated service subsample at Lirhembe, cooperative members adopted special agricultural promotions differently, depending on the number of services they were exposed to. Initially, a cooperative network facilitated the production and marketing of European vegetables, such as cabbages, tomatoes, and onions; a Lirhembe-owned vehicle (destroyed in an accident a year after the program started) transported vegetables to nearby schools, markets, and Kakamega township at minimal costs to farmers. Mexican 142 hybrid beans were another promotion which required purchased seeds and special labor.

**Table 18. Responsiveness to Two Crop Promotions by Farm Management Type, Controlled for Different Amounts of Services Extended (Lirhembe Sample Only)**

| | Low servicing (0-2 services) | | High servicing (3 or more services) | |
|---|---|---|---|---|
| | Female | Joint | Female | Joint |
| Grew Mexican 142 beans | -- | 30%/6 | 50%/5 | 31%/8 |
| | N=28 | | N=36 | |
| Grew European vegetables | 25%/2 | 50%/10 | 78%/7 | 48%/13 |
| | N=28 | | N=36 | |

Unlike Shikulu female farm managers, who operated effectively without administrative aid, Lirhembe female managers did not perform as well with few services compared to jointly managed farms, as evidenced in adoption rates. In the high service category, however, female managers respond dramatically to service relative to the negligible response of jointly managed farms. Besides suggesting the improved performance which occurs with administrative investments in female-managed farms, the data also lend support to administrative rather than land scarcity explanations for declining productivity. It appears that female managers, because of their organizational resources, have a performance edge in context with minimal services, but that in more intensively administered contexts, the balance is tipped in favor of farms where a man is present until saturation services reverse the situation again for certain crops. A key question, then, is whether policy should aim solely at linking farmers' performance to costly administrative support, or whether efforts could be directed toward restoring once-innovative women's organizations with their labor and information sharing with appropriate crop mixes.

# AN INQUIRY INTO FACTORS UNDERLYING GENDER DIFFERENTIALS IN AGRICULTURAL POLICY

## ALTERNATIVE EXPLANATIONS OF FINDINGS

Clear differentiation in the delivery of agricultural services to jointly managed and female-managed farms has been documented. In each service examined, discrimination against female managers was found, except in the brief provision of tractor services to Lirhembe cooperative members. The gaps increased as the value of the service increased, from the rare demonstration plots and home visits to the very aggravated gaps in specialized training and in credit. Saturating services in an area only reduced inequity when services were divorced from political influence, but politics pervaded the provisioning of services. These inequitable patterns cast doubt on the government's ability to realize policy objectives.

By way of summary, the information below contrasts mean numbers of services for a cumulative service scale in the ordinary, intensified, and saturated areas. Service items were simply added. 31

Table 19. Mean Services by Administrative and Farm Management Type

| | Female-managed farms | Jointly managed farms |
|---|---|---|
| Ordinary:<br>Shikulu (N=108) | .8 | 1.6 |
| Intensified:<br>Shitoli (N=104) | 1.6 | 2.2 |
| Saturated:<br>Lirhembe cooperative (N= 65) | 3.6 | 4.0 |

Predictably, the more intensified the services, the higher the mean number of services received. Yet increases for female-managed farms lag behind jointly managed farms in all settings. Quite strikingly, residence in intensified settings merely raised female managers' access to the level of jointly managed farms in ordinary areas. Yet declining service differentials -- 0.8 in the ordinary area, 0.6 in the intensified setting, and 0.4 in the saturated setting -- represent narrowing gaps with more administrative resources. With the extraordinary staff and support services, however, one would have expected the gap to be considerably narrowed or eliminated. Extraordinary levels of service seem necessary to compensate for the more typical sex discrimination found in ordinarily served sublocations. This slight narrowing of the differential is achieved at the cost of farmers' dependency on a costly and ultimately unreliable administration, particularly when external resources are involved, such as in Lirhembe. Three explanations are set forth to explain differentials.

## EXPLANATION 1: PREJUDICIAL ATTITUDES AND IDEOLOGICAL BIAS

Although administrative field staff attitudes were not a prime focus of this research, the elimination of income, land size, and performance of differentials suggests that the most important factor in explaining differentiation are prejudicial attitudes and ideological bias. A number of agricultural instructors who were interviewed made prejudicial comments about women farmers. More important, though, was their expressed avoidance of women because of customary patterns whereby men speak to men and women to women.

> In the African way, we speak to the man who is the head of the house and assume he will pass on the information to other household members. Being men, of course, it is easier for us to persuade men.

Previous sections have documented the preferences of technical staff for speaking with men.

Rather than rely on staff attitudes, the sample data allow a test of alternative explanations for bias against women, based on actual staff behavior. It is not unreasonable to assume that staff, facing time and resource constraints, would concentrate on the better off (who have more cash to purchase seed, fertilizer, and labor), on the owners of larger plots (who have more land on which to experiment), and on the high performance, innovative farmers (who would make officers look as if they achieved goals). If none of these factors explain the differences in service delivery, then the impact of prejudicial attitudes on women's productivity becomes more apparent.

### Income

A common finding in extension studies is a tendency for farmers' access to services to increase with wealth. As shown earlier, though, female-managed farms span all economic categories and have access to various cash-earning activities as well as husbands' salaries. The economic standing scale was dichotomized into low and high, with low representing near subsistence living and very little access to cash.

In comparing visits by agricultural instructors, 57 percent of low-income female managers had never been visited, in contrast to only 39 percent of jointly managed low income farms. That poor women receive the fewest visits is fairly predictable. More surprising is how the bias against women was maintained at high-income levels. A full 39 percent of high-income female-managed farms never received a visit, compared to just 16 percent of the jointly managed high-income farms.

This same tendency is borne out in looking at the mean number of services received on a summary scale. Low-income female-managers received an average of 1.1 services, while their wealthier sisters received 1.4. In contrast, low-income farms with a man present received an average of 1.5, but their wealthier couterparts, 2.4. Poor jointly managed farms did better than wealthier female-managed farms. The differences between low-and high-income female-managed farms are not so marked, suggesting that the consequences of agricultural policy for women managers as a group are not as differentiated as for jointly managed farms.

## Land

Female managed farms also span all land sizes, as they are holding land for migrant husbands who will return upon retirement. There is a slight tendency for female managers to inflate low acreage because approximately a quarter of them are widows. Yet widows are, in fact, slightly more likely to get services because of their age and the prestige that it brings, their length of time in the community, and the reduction of suspicion about unrelated men (staff) and women together. In an analysis of those farms over five acres, numbering only 23 households (and thus demanding cautious interpretation), 38 percent of female-managed farms had never been visited by an extension officer, compared with ony 7 percent of jointly managed farms. Land, like income, does not hold as an explanation for differences in service delivery.

## Farm Performance

Extension officers understandably seek responsive clients who are willing to diversify and innovate. In the table below, a summary service scale allows an examination of the relationship between management type and service access, comparing farms with one versus two or more cash crop innovations.

Table 20. Cumulative Service Scale by Farm Management Type Controlled for Number of Cash Crops Grown.

| | 0-1 Cash Crop | | 2+ Cash Crops | |
|---|---|---|---|---|
| | Female | Joint | Female | Joint |
| 2 services | 92%/46 | 72%/50 | 63%/22 | 50%/29 |
| 3 services | 8%/4 | 22%/15 | 34%/12 | 29%/17 |
| 4-5 services | -- | 6%/4 | 3%/1 | 21%/12 |
| | N=119 | | N=93 | |

Female managers who diversify cash crop production are clustered in the low-service categories. Among farms with two or more cash crops, slightly over a third of the female managers had access to a broad variety of services, in contrast to half of the jointly managed farms. However, even in that third, only a minute proportion of female managers (only one case, in fact) had four or more services extended to them. High-performance female managed farms were not much better served than jointly managed farms that showed little adaptiveness. Where only one cash crop or less was grown, a quarter of the jointly managed farms had access to three or more services, which was true for only 1/12 of comparable female-managed farms. Extension staff did not seem to avoid poor performance farm households, as long as a man was present.

In an examination of *early adopters*, that is, the 50 farmers who grew hybrid maize as early as five or more years before the research, almost a third of female managers, or 31 percent, had no administrative support or advice for such a move, while only 3 percent of the jointly managed farms were so neglected. Timely adaptiveness, just as crop diversification, actually underestimates female managers' capabilities, given the labor constraints of female managers.

Female managers' performance as measured by their readiness to innovate and by productivity has been documented in other studies about Kenyan agriculture. Peter Moock found that women's maize yields per acre in Vihiga Division (south of this study's research site) were similar to men's, and even surpassed men's when access to education and extension was controlled. Similarly, John de Wilde noted "quite a few women" among progressive farmers, as did studies in neighboring Muganda (see note 13).

## EXPLANATION 2: WOMEN'S INVISIBILITY

Women are conceptually and politically invisible to the administration. They rarely attend barazas, and the continuing notion of male headship over households, however symbolic given extensive male absence, precludes full attention to women. But women's behavior, both as individuals and collectively, also is part of the explanation for their invisiblity. Women rely on their own labor and mutual aid groups, rather than on government. And women's alternative institutions that compensate for non-access to services appear to operate as well as those of the agricultural administration except in Shitoli where intensified and saturated services increase overall dependency on government. Yet women's groups are invisible to the administration.

Only rarely do farmers initiate individual contact with agricultural instructors, and very few women managers are among those who do. In the full sample, only 50 had ever initiated contact, and seven were women from female-managed farms. Among female managed farms, 88 percent never initiated contact, in contrast to 68 percent of jointly managed farms. Women seldom make such contacts because they fear that requesting special attention induces suspicion and incurs obligations. Women have traditionally depended on themselves and other women in the agricultural sphere, a pattern reinforced by program implementation and male personnel.

Agricultural administration is ensnarled in local politics, and those with political influence are likely to derive benefits. Only infrequently have women -- largely elite women -- banded together politically and become more visible. And the way they have defined women's issues has not helped female managers and farmers. For example, during the 1972 M.P.'s campaign, wealthier women approached him as a delegation to request services for women. The manner in which services were requested ranged from a very diffuse request that the M.P. "remember the work women do" and "how they need help," to more specific requests for donations for church activities, a nursery or women's center. These women delegates, from the local elite and well educated for the area, were past recipients of domestically oriented government programs at five times the rate of ordinary women, programs that included marriage training and "homecraft" such as cooking, sewing, and knitting. Many were schooled in the postwar era, when female education emphasized homemaking skills. Given their training, elite women tend to define "women's issues" within the realm of home-, child-, and church-related areas, rather than take a more comprehensive focus on income-earning activities derived from agriculture.

Owing to male out-migration, women are a majority of the electorate in the M.P.'s electoral constituency. Though clan and subclan groups, which coincide with residential areas, receive priority attention in campaign strategies, blocs of women that cut across those key groupings are also important. The M.P. needed women's electoral support, and women were aware of the potential profitability of linking themselves with him. However, their political visibility resulted in the creation of a women's sewing and knitting class in the cooperative staffed by a home economics assistant. While the cooperative did benefit those women who joined, this domestic-oriented service was enjoyed by only a tenth of the women in the sublocation who had time.

Unless they become more visible, individually and collectively, women will be by-passed in agricultural administration. But it is important to consider which women are visible, what interests they represent, and how effective their strategies are. In a brief, two-year episode of political integration in the late colonial era, Chief William mobilized women on a location-wide basis to use their labor for land consolidation efforts and to expand his political base of support. Women were granted some authority in the form of "women subchiefs" who represented women in barazas. Women developed a complex organization called Umoja, large-scale income generation through collective agricultural labor, and even women's courts, where women judges mediated infractions among women. 32 Women's brief integration through Umoja in the late colonial political structure coincided with administrative outreach to women's groups. Elite women's contemporary political activity does not.

## EXPLANATION 3: ADMINISTRATIVE CONVENIENCE

Common in bureaucracy is the convenience of continuing earlier policies and only reforming them incrementally. The wholesale importation of older Western notions about men as 'appropriate' farm managers and women as 'appropriate' homemakers, though convenient, was in error and ignored the facts. Following from these assumptions was the tendency to separate farm clientele administratively and to segregate staff by sex, creating two obvious problems: the minute proportion of women field staff--2 percent in Kenya (3 percent for Africa as a whole), 33 and the dilution of agricultural content. Kenya's experience bears out Uma Lele's observation that "the goal of extension services has frequently been not the increase in farm level productivity of women but rather finding ways to reduce their participation through promotion of more homebound activities." 34

The neglect of female managers and small farmers generally is sometimes dismissed as the result of cash crop rather than food crop emphasis. However, maize has doubled as a food and cash crop promotion since World War II. Still, female maize producers experience bias. One wonders how much more aggravated bias might have been if purely cash crops were the focus.

The conveniences of staff and of using existing procedures also forestall attention to all farmers. Male field staff find it convenient to speak with men, to use men's communication vehicles such as barazas, and to rely on elite farmers (some of whom are women), for their sporadic demonstration plots. Loan staff also find it convenient to rely on title deeds as the security for loans. These strategies born of inertia result in serious inequities and an underutilization of existing resources.

# A DISCUSSION OF ALTERNATIVE IMPLEMENTATION POLICIES

Merely increasing or intensifying services would have little effect on addressing ideological bias, women's invisibility, or inappropriate administrative conveniences. The Kenya government's official reasons for constraints in agriculture--knowledge, technology, and credit 35 --are only a small part of the picture. Beyond this, a gender-specific approach is needed which recognizes gender disparities in access to and control over resources and gender-specific communication patterns. Women producers must be viewed as legitimate agricultural modernizers. In practice, general neutrality perpetuates differentiation and ultimately, ineffective implementation. Home economics, which does not view women as managers or producers, cannot substitute for transforming the agricultural administration to serve its key clientele.

The suitability of "more of the same," or simply intensifying or saturating an area with services, is highly questionable. Added extension staff, while they improved the likelihood of home visits, did not do so proportional to their numbers. More staff operate in the same type of administrative structure, which provides few incentives for effective or equitable administrative performance for farmers. The costs of continually expanding permanent staff are too great in resource-scarce contexts like Kenya. Similarly, the costs of saturating services are virtually out of the question for the whole rural countryside. Lirhembe, while innovative and redistributive, was near collapse due to loan defaults, grade cow death from tick diseases, and the breakdown of the tractor and vehicle, too costly and sophisticated to repair locally. But it is not merely government budgets that bear the cost of administrative intensification. Another high cost of intensification is the generation of farmers' dependency on government to increase productivity rather than on their own indigenous resources.

Planners in countries like Kenya should consider alternatives which range from improved management and supervision to reduced reliance on extension officers and their home visits in ways which spread responsibility among men and women professionals, paraprofessionals, and community groups. Strengthening existing women's groups would build on indigenous resources to permit information and labor sharing, economies of scale, and guarantees for loans. These groups, broadly representative of all women, should also link with the local political process.

The details of these strategies--virtually cost-free save administrative inertia--are outlined below and summarized in Box 1, in their various components: staff and staff training, extension and farmer training, credit, and cooperatives. Of course jogging administrative inertia is no easy matter, involving as it does institutionalized bias and existing procedures. Administrative will and leadership are essential. Extra funds may be required to put procedures in place; or a redistribution of funds inside agencies may be necessary. However, the costs involved are far outweighed by the elimination of wasteful expenditures and ineffective procedures as well as by the gains resulting from the changes.

The alternative strategies below, designed mainly to affect the point of contact between field workers and farmers, should be seen as accumulating from improved management procedures to slight reforms in the administrative structure, to a more thoroughgoing transformation of that structure which involves more considered shifts of policy. They do not address the appropriateness of various service packages, which bear ongoing examination, but rather to examine what options for extending benefits more evenly and equitably exist at different levels of change. Furthermore, the recommendations for not restricted to Kenya; some of them could be generalized to countries with similar situations.

## ALTERNATIVE A: IMPROVED MANAGEMENT

Suggestions offered in this section do not require an increase in staff or changes in administrative structure. They primarily relate to field extension methodology and the more efficient utilization of available resources. 36

### Staff and Staff Training

Kenyan agricultural instructors work in an administrative structure which gives them little incentive to perform effectively, efficiently, or equitably. An improved management strategy would alter the composition of field staff and limit their discretion by establishing procedures which direct field workers to serve those now disadvantaged while at the same time making staff performance matter in an incentive structure which rewards or penalizes specific behavior.

Presently, there are gross imbalances in the gender composition of staff. Because male staffs' avoidance of women managers is a primary reason for the inequities outlined in this case, more female staff should alleviate this problem.

In addition, in most cases female recruits are tracked primarily into home economics, which neither aims to serve women as farmers nor has the budget and staff to reach more than a minuscule percentage of "farm wives." Male extension staff generally receive little to no home economics training, the content of which carries over into agriculture. 37

Administrative leaders should consider establishing targets for minimum proportions of women in agriculture staff training, along with extra cash incentives, if necessary, to recruits and to the colleges, to alter exclusionary practices of the past. More female staff in the context of a majority female farm clientele and gender-specific communication patterns will increase the likelihood of serving women and women managers. Management might also consider retraining home economists, who already receive some agricultural training, although this move would deplete an already sparsely staffed division if it is to retained.

Worries are frequently expressed that female staff would not accept postings away from families. Management should consider the importance of stationing staff--male or female--near their families and furthermore, increasing the stability of that placememt. As it is, transportation problems for agricultural instructors limit their time with farmers, and frequent staff rotation reduces instructors' accountability for change within given areas.

Still, the overwhelmingly male staff is responsible for serving women managers, for which special incentives will be necessary. Management should institute a performance appraisal system which rewards effective and equitable staff performance. For example, a monitoring system could be established to assure equitable performance. Or targets could be set to increase contact with disadvantaged clientele, such as women managers, along with monitoring mechanisms to document the realization of goals. Staff who do not meet goals would receive poor evaluations with consequences for their salary increases and mobility prospects.

Such a reform might add some paperwork to an already overburdened system. But these monitoring mechanisms could be easily mainstreamed into existing reporting procedures. For example, a system could be devised to review reports on who is visited and ensure that men and women are visited proportionately. Moving beyond the penalties implied in such procedures, staff might also be rewarded monetarily, with bonuses for equitable delivery. Such a reform would be cost-free if the increments of upcoming salary increases formed the fund for this two-tiered payment system, or if status--in the form of awards or titles--constituted the reward.

Of course, both new and current staff should be aware of gender inequities, the legitimacy of women producers, and the desirability of increasing their productivity in new and in-service training programs. The trainers themselves must be exposed to and encouraged to incorporate this type of information in the curriculum. Such information goes beyond offering a home-economics course option to trainees.

## Extension and Farmer Training

Home extension allows individual, personal contact between farmers and field specialists, adapted to unique farm conditions. But it is a costly means to deliver information, and the discretion available to field workers makes rife the possibilities of both preferential and discriminatory treatment.

Group extension increases the prospects of accurate, relevant, and visually convincing information transmission with potentially more equitable outcomes. Problematic, though, are the communication channels used to announce demonstration plots and the groups for whom demonstrations are made.

Administrators may consider promoting the use of more group extension, after first assuring that staff understand the community social structure. Those staff must carefully consider the kinds of communication channels in which messages are transmitted to assure that dissemination occurs through women's networks as well as men's. The focus should shift away from individual home visits; group extension, such as demonstration plots and continuous work with local groups, should be used with both separate women's groups and men's groups, or integrated groups, depending on community structure. In many cases, this involves identifying existing neighborhood groups both for dissemination and demonstration.[38]

Farmer training provides more intense instruction on specialized topics. Under the current approach, training staff are competent and monitored effectively. Yet farmers perceive training as available "by invitation only," and hence as something to confer status. Farmers must be assured that any and all are entitled to receive training.

### Credit

Credit is virtually unattainable for all but a select few. Were low-risk, in-kind loans, such as the Guaranteed Minimum Return, to have a reduced land minimum, a larger number of farmers should have access. All credit recommendations add cost to the extent that access expands among small farmers.

Institutional procedures for acquiring credit make it extremely difficult for women managers to obtain capital from this source. Policymakers should consider adopting minimum quotas to assure that women farm managers receive AFC loans.

### Cooperatives

Most cooperative recruitment is aimed at men, and consequently, women managers or wives either do not join or they use benefits less frequently than they are entitled to. Lingering but increasingly irrelevant notions of "male heads," regardless of where they are located or what they contribute, reduce the likelihood that women will seek membership or that there will be cooperative out-reach to women.

In some cases, one could recommend recruitment of men and women as an improved management procedure. In the case of Kenya, however, cooperative laws which require members to hold land title deeds perpetuate bias against women, given the way that land reform has excluded women from formal ownership.

Systematic recruitment of women farm managers, based on their producer status rather than land-ownership status should be encouraged. Also, when partners live together, joint membership in cooperatives should be required to encourage both women's participation and their visibility. Along with that, cooperative benefits and crop payments should be allocated directly to producers, including women. Life staff training, information about gender inequities and the legitimacy of women producers should be incorporated into cooperative training.

## ALTERNATIVE B: INTERMEDIATE STRUCTURAL CHANGE

An intermediate change strategy goes further than simple improved management to reform structures slightly by expanding or changing the composition of operating staff, the manner of disseminating services, and by promoting minor shifts in policy. This does not necessarily imply greater cost, only more energy in reform efforts. Alternative B

## BOX 1. ALTERNATIVE STRATEGIES TO REACH WOMEN FARM MANAGERS

| ALTERNATIVES | AREA OF CHANGE | RECOMMENDATION | RATIONALE | ADDED INFORMATION REQUIREMENTS | DIRECTION OF INCENTIVES | COMPLEXITY OF CHANGE | COST |
|---|---|---|---|---|---|---|---|
| A MINIMAL | STAFF AND STAFF TRAINING | Establish targets for minimum proportions of women into agricultural training. | Male staff avoid women farm managers. | None | Top-down | Low | Low (unless financial incentives utilized) |
| | | Increase stability of staff placement. | Frequently transferred staff lack specific area expertise and accountability for results. | None | Top-down | Low | None |
| | | Recognize desirability of both men and women staff to be stationed near families. | High staff turnover decreases quality of performance; distant transfers the rationale for low female recruitment. | None | Top-down | Some | None |
| | | Devise incentives for male staff to serve women farm managers, such as MONITORING to assure equitable performance and/or TARGETS to increase contact with disadvantaged clientele. | Overwhelmingly male staff avoid certain farm managers and lack incentives to serve them. | None, but monitoring procedures. | Top-down | Some | Low - (Administrative) |
| | | Reward staff monetarily for equitable delivery. | Overwhelmingly male staff avoid certain farm managers and lack incentives to serve them. | None, but monitoring procedures. | Top-down | Some | None (If salary increments the funding source) |
| | | Incorporated information about women producers and inequities into training. | Staff lack cues to improve equitable performance. | Some, for trainers. | Top-down | Low | None |
| | EXTENSION STRATEGY | Promote group extension with women's and men's groups. | Group outreach extends quality outreach in equitable ways. | Some, community groups and communication networks. | Top-down | Low | None |

| | | | | | | |
|---|---|---|---|---|---|---|
| | | Disseminate information about training through women's and men's networks. | Information reaches men primarily. | None | Top-down | Low | None |
| | | Avoid substituting home economics training for agriculture. | Home economics dilutes full agricultural training. | None | Top-down | Low | None |
| | CREDIT | Reduce land minimum requirement for Guaranteed Minimum Return loans. | Small farmers cannot qualify now. | None | Top-down | Low | Some, associated with expanded credit. |
| | | Establish quotas to assure that women farm managers receive AFC loans. | Women farm managers rarely get loans. | None | Top-down | Low | None |
| | CO-OPERATIVES | Recruit women farm managers. | Women not tapped in recruitment. | None | Top-down | Low | None |
| | | Require joint membership for partners living together. | Recognizes women as de facto producers, managers, members, and representatives. | None | Top-down | Low | None |
| | | Allocate co-op benefits and payments directly. | Benefits serve as incentives for actual producers. | Some | Top-down | Some | None |
| | | Incorporate information about women producers and inequities in training. | Co-op managers lack cues about problems. | Some for trainers. | Top-down | Low | None |

| ALTERNATIVES | AREA OF CHANGE | RECOMMENDATION | RATIONALE | ADDED INFORMATION REQUIREMENTS | DIRECTION OF INCENTIVES | COMPLEXITY OF CHANGE | COST |
|---|---|---|---|---|---|---|---|
| **B** INTERMEDIATE (all of the above PLUS) | STAFF AND STAFF TRAINING | Create network of paid, part-time paraprofessionals, both men and women. | Professional staff cannot adequately cover current territory. | Some -identify farm leaders | Top-down | Initially, some | Some |

## BOX 1. ALTERNATIVE STRATEGIES TO REACH WOMEN FARM MANAGERS (CONT'D)

| | | | | | | | |
|---|---|---|---|---|---|---|---|
| | | Reward professional staff monetarily from community revenues. | Staff avoid certain farmers; lack incentives to serve them. | Some | Bottom-up | Initially, some | Cost to community |
| | | Solicit evaluation feedback from community committees on staff performance. | Staff not accountable to communities they serve. | Some -appropriate representative committees. | Bottom-up | Initially, some | None |
| | EXTENSION | Bring training staff to the community. | Distant training difficult for women farm managers to attend. | None | Top-down | Some | Some travel |
| | CREDIT | Group loans to producers lacking title deeds or salaries. | Procedures exclude many producers, such as women managers. | Some -existing groups or group formation. | Top-down | Some | Some associated with expanded credit. |
| | | Lending facilities located closer to farmers (or mobile units). | Less mobile farmers excluded. | None | Top-down | Initially, some | Some transport |
| | CO-OPERATIVES | Make co-op management and participants reflect community producer composition, such that half or more women. | Currently unrepresentative. | Some, community composition. | Top-down | Initially, some | None |
| C THOROUGH (all of the above PLUS) | STAFF AND STAFF TRAINING | Require staff to work with paraprofessionals and representative community members to establish collaborative goals and strategies for agricultural change. | Staff not accountable to community. | Some, representative community members. | Bottom-up | Initially, some | None |

| | | | | | | |
|---|---|---|---|---|---|---|
| | Integrate agricultural and home economics staff. | Segregated staff tend to segregate farm community. | None | Top-down | Very complex | Initially high |
| EXTENSION | Develop a "Farming Systems" approach to agricultural research with farmers. | Research not locally adapted or judged by farmers' criteria. | None (would augment research with local applications). | Bottom-up | Initially, some | Some -training of staff |
| CREDIT | Establish joint title for land ownership with corresponding names on title deeds. | Procedures exclude women producers from credit and from control over assets. | Some, spouse names | Top-down | Very - involves other ministries | High, initially |
| CO-OPERATIVES | | | | | | |

should be viewed as an add-on to the cumulative set of changes recommended for Alternative A. Its effectiveness depends on completing the above changes.

### Staff

Besides those changes recommended for staff above, management should consider ways to extend quality staff coverage even further with a network of paid, part-time paraprofessionals, both men and women who work with neighbors in areas coverable by foot. Already, innovations are diffused through informal means, but the quality of technical information being passed is uncertain. Trained paraprofessionals would improve the quality of that information in systematic ways. Local paraprofessionals are familiar in the community, and their residence in the area would enhance interaction with farmers. Women could be easily recruited. Besides, the lower salaries of men and women paraprofessionals, compared with those of professionals, would represent cost savings compared with simply augmenting professional staff.

To increase the equitable performance of professional staff while at the same time broadening their accountability to men and women in the communities they serve, the upper tier of monetary rewards for their work along the lines of the two-tiered system described above would be derived from community sources. Such a reform, however, has the danger of politicizing delivery even more to men and to the already privileged, to the extent community participation structures are unrepresentative. Therefore, increasing women's political participation to produce accountability to women is essential. Reducing the complications of that recommendation somewhat but at the same time meeting its rationale would be the strategy of soliciting evaluation feedback from representatives, male and female, from the communities in which agriculture staff work.

### Extension and Farmer Training

Physical distance between farmers and training centers, requiring days or weeks away from the residence, limits farmer trainees to those households with at least two adults. Furthermore, cultural constraints on women's mobility may make female participation virtually impossible. Policymakers should create training alternatives, such as bringing training staff to the community to run comparable programs out of a school, church, or community center. Travel expenses would add to costs, but the result would be widely expanded access for farmers. Currently, training staff are underutilized, so such a reform would improve the effectiveness of their professional time. Again, trainers must recognize that home economics is no substitute for agricultural training.

### Credit

Besides expanding the coverage of current loans through existing, but improved procedures as outlined above, policymakers should develop alternative means to deliver and secure the payment of credit. For example, loans might be extended to existing or newly formed women's groups, secured through group liability and consequent peer pressure to guarantee repayment. Also, administrators should consider locating lending facilities closer to farmers, especially for those less mobile, such as women managers.

### Cooperatives

If women are to be recruited into cooperatives effectively, (assuming a policy framework which does not limit cooperative membership to landholders), steps should be taken to ensure that women are represented equitably in management, to reflect their numbers in the community. This should also increase the number of women who attend cooperative training programs which, like farmer training, might best be located in the community.

## ALTERNATIVE C: THOROUGH STRUCTURAL CHANGE

A more thorough change goes furthest to alter management and their structures in more fundamental ways which involve the community as well as the administration, and involve broader policy decisions. These changes work to empower individuals and involve them in ongoing flexible program goal definitions.[39] Again, the following recommendations will only be effective if alternatives A and B are pursued.

### Staff

Once properly supervised and made more accountable to the community, staff should be required to work with paraprofessionals and representative farmers, reflective of numbers in the community, to establish goals and strageties for agricultural change. Women would be active participants. Communities would share responsibility with the administration over strategies for change that affect them.

## Extension

Rather than separating research from extension and from farmers, experimental research should be conducted with farmers on location and should be judged according to criteria farmers use when making adoption decisions, such as risk reduction, appropriate crop mixes, and improved income. Appropriate here is a farming systems approach which takes into account different management structures, female and otherwise, labor constraints of women and men, and differential access to inputs and land.

Policymakers should also consider integrating agriculture into home economics staff training--already accomplished to a certain extent--and also integrating home economics into agricultural training with the ultimate aim of a single agricultural extension service, with home economics as a district level specialization. A bifurcation of services by task, gender, function, and clientele is unrealistic, inefficient, and ineffective.

## Credit

The fundamental cause for women managers' low access to loans is land reform which excluded women producers from control over and title to land they till. As such, they operate much like tenants lacking secure access to and stake in this productive resource. Steps should be taken to establish women's ownership of land, either jointly with spouses or as individuals, with corresponding names on title deeds.

## General Policy Change

Besides these specific recommendations, other more general policies relating to price and participation which now hinder small farmers and women managers should be reconsidered.

A strategy less dependent on bureaucracy, and thus less costly, would improve guaranteed prices, now too low in some cases to cover input costs or to provide incentives for farmers to produce more. Low prices for staple goods frequently placate urban consumers rather than make higher production attractive for farmers. And producers, women included, must be compensated directly for their work to provide incentives. Moreover, either market restrictions should be lifted, or the number of maize bags that can be moved across district boundaries should be raised. Among small farmers in Kenya, these restrictions fall heavily on women traders. In the best scenario, they hide maize and pass through police checkpoints, pay the police to pass with theAir goods, or, at worst, have their maize confiscated.

Women, both individually and collectively, must be integrated into local community politics and development committees in order to learn of and receive the many politically charged benefits. Kenya's official desire to increase rural participation must extend to women in balanced ways, in proportion to their numbers and developmental importance. Separate women's bureaus must also see their role as making all policies sensitive to women rather than merely social services, community development, or traditionally separate women's activities. To ensure this policy integration, women's advocates, with resources and authority to both encourage and penalize, should be positioned in national, district, and local offices to work with women's groups toward this aim.

The recommendations in Alternatives A through C move from the more simple, management-oriented changes to those that involve the community, necessarily more complex. Yet such community involvement, besides jibing with Kenya's goals to increase participation, reflect a more people-oriented strategy which builds both capacity among community residents and accountability from staff, and thus a more holistic development. Virtually all of the recommendations are cost-free, save the energy necessary to institute change and overcome the usual bureaucratic inertia. Taken alone, each would produce outcomes that improve the effectiveness of agricultural administration. Adopted together, the recommendations should have a profound impact on addressing the issues raised in this case study.

## CONCLUSIONS

Historically, women in western Kenya have produced food for the community and contributed considerable resources to households. With the introduction of new economic and political forms, agriculture was increasingly commercialized, employment was made available to men earlier and more extensively than women and agricultural advice, training, and loans were oriented primarily to men. An idealized view of the family was introduced and diffused through government programs in which husbands are breadwinners and women are relegated to the domestic rather than productive sphere. Still, women are active in agriculture, as workers and managers, and they belong to thriving associations. Yet early agricultural policy was explicitly tied to idealized assumptions about the family. Agricultural services were initially and predominantly delivered to men, a result of administrative structures largely comprised of men who implemented policy in a setting with strong gender-specific communication patterns.

Governments both before and since the seventies have channelled common agricultural services--such as agricultural instructors' visits--to men, as well as the most valuable services such training and loans, the latter being tied to the control of other resources such as land and wages. In fact, disparities between female-managed farms and other types of farms widen as the value of the service increases, both in cost to the administration and in benefit to the farmer. Expensive strategies such as intensifying or saturating areas with services do little to alter these disparities. A bias against female managers exists, independent of other factors such as economic standing, land size, and performance. Indeed, wealthy and innovative women farming on relatively large tracts of land experience discrimination when compared to their male counterparts. Land reform has reinforced this pattern, making it difficult for women managing farms to acquire loans.

Despite Kenya's well-developed extension system, a pervasive bias exists against an important and large part of its clientele, women managers. The administrative failure to reach this large female clientele is a telling critique on government commitment and ability to affect agricultural change. By pursuing a range of low-cost administrative reforms which recognize and build on indigenous capability, including those of female managers, Kenya has the potential to increase productivity and thereby ameliorate the domestic food crisis.

# ENDNOTES

1 The policy objectives came from two sources. First, objectives are laid out in the Government of Kenya, *Development Plan for the Period 1974 to 1978* (the period of this research), Part I, (Nairobi, Government Printer, 1974), pp. 18ff and Chapter 10 on Agriculture, Land Settlement, and Cooperatives. Second, objectives are inferred from program implementation in the field research area.

That Kenya is concerned with equitable development can be derived from frequent reference to the goals of "social justice," a "more equitable distribution of income," and the "equitable sharing of its (economic expansion) benefits." pp. iv., 1, *Development Plan*. The plan also makes reference to the Sessional Paper no. 10, "African Socialism and Its Application to Planning in Kenya," 1965, a development-with equity philosophical statement which this plan continues.

2 The quote is from Louise Fortmann, "Women's Work in a Communal Setting: The Tanzanian Policy of Ujamaa," *Women and Work in Africa*, Edna Bay, ed. (Boulder: Westview Press, 1982), p. 195. See also, Ester Boserup, *Woman's Role in Economic Development* (New York: St. Martin's Press, 1970); Jette Bukhe, *Village Women in Ghana* (Uppsala: Scandinavian Institute of African Studies, 1979).

3 This and the following comments from Wagner are found in Gunter Wagner, *The Bantu of North Kavirondo*, Vol. 1 (London: Oxford University Press, 1949) pp. 41, 20, 45, 156. Kavirondo is the former name of Kakamega. Wagner's work was done among the northernmost Bukusu and southernmost Maragoli subgroups of the Balyuia. Idakho location is just north of Maragoli location. See also another anthropological study by Walter Sangree, which focuses on the Luyia group in the southeastern part of the district, *Age, Prayer & Politics in Tiriki*, Kenya (London: Oxford University Press, 1966). Also of interest is a government time allocation study entitled "A Report on Economic Studies of Farming in Nyanza Province, 1963," Farm Economic Report No. 26, 1969.

4 Kenya National Archives, Native Agricultural File, Center and North Kavirondo(1930-46).

5 Judith Heyer. *Survey of Agricultural Development in the Small Scale Farm Areas of Kenya since the 1920's* (Nairobi: University of Nairobi, Institute for Development Studies) Working Paper No. 194, October, 1974, pp. 15-16.

6 Kenya Colony & Protectorate, *A Plan to Intensify the Development of African Agriculture in Kenya*, Compiled by R.J.M Swynnerton (Nairobi: Government Printer, 1954), p.53.

7 Swynnerton, p. 10.

8 Everett Rogers, *Modernization Among Peasants; The Impact Of Communication* and *The Diffusion Of Innovations* (New York: Free Press, 1969, 1962).

9 Judith Heyer, D. Ireri, and Jon Moris, *Rural Development in Kenya*, (Nairobi: East African Publishing House, 1971); selections in James Sheffield, ed., *Education, Employment and Rural Development* (Nairobi: East African Publishing House, 1967); *ILO/UNDP, Employment, Incomes and Equality: A Strategy for Increasing Employment in Kenya* (Geneva: ILO, 1972); Joseph Ascroft, Fred Chege, Joseph Kariuki, Niels Roling, and George Ruigu, "Does Extension Create Poverty in Kenya?" *East Africa Journal*, 1972; David K. Leonard, "Why Do Kenya's Agricultural Extension Services Favor the Rich Farmers?" Paper presented at the 16th Annual Meeting of the African Studies Association, Syracuse, New York, October-November 1974.

10 This and following historical evidence came from the Kenya National Archives. Nyanza Province. Central and North Kavirondo District *Annual Report*, 1948; Native Agricultural Files.

11 Information in this paragraph comes from the following documents: North Kavirondo District *Annual Report*, 1951, p. 14, and 1952, pp. 45, 70; Special supplement to the District *Annual Report*, 1955, p. 11: District *Annual Report*, 1959, pp. 26, 28. The Kenya Colony and Protectorate *Annual Reports* also report the recruitment of women agricultural assistants, as does Lord Hailey, *An African Survey* (London: Oxford University Press, 1956), who reports that 48 female staff were employed in agriculture, p. 894. The Ikolomani Division, Miscellaneous Affairs file, 1958-59, reports that district-level male agricultural staff had a parallel female staff.

12 Bertha F. Strange, *The Story of Kenya's Home Extension Service: 1962-65*, (Washington, D.C.: U.S.A.I.D. Mission to Kenya, 1965), p. 1.

13 See Peter Moock, "Managerial Ability in Small Farm Production: An Analysis of Maize Yields in the Vihiga Division of Kenya," (Unpublished Ph.D. Dissertation, Columbia University, 1973), p. 168; and his "The Efficiency of Women at Farm Managers: Kenya," *American Journal of Agricultural Economics* 58, 5, December, 1976. See also John de Wilde, *Experiences with Agricultural Development in Tropical Africa* (Baltimore: Johns Hopkins University Press, 1967), p. 169; Edgar Bowe and Jon Moris, "Social Characteristics of Progressive Buganda Farmers," *East African Journal of Rural Development* 2, 1,(1969), p. 57; Judith Heyer and Joseph Ascroft, "The Adoption of Modern Practises on Farms in Kenya: Preliminary Results of a 1968 Survey of Farms across Kenya," (Dar es Salaam: University of East Africa Social Science Conference, Volume V, December

27-31, 1970), p. 336.

14 Republic of Kenya, "Report of the Agricultural Education Commission," Chaired by Weir (Nairobi: Government Printer, 1967), p. 13, contains the information on the number of women officers. Those figures matched the 2% figure in Kakamega District.

15 All demographic, spatial, and density information in this study comes from the Republic of Kenya, *Kenya Population Census*, 1969, Vols 1-3 (1970), unless otherwise indicated.

16 S.H. Ominde refers to the uneven sex ratio in his *Land and Population Movements in Kenya* (Nairobi: Heinemann, 1968), p. 109. Wagner provides male out-migration figures for the 1930s, II, p. 94, and Judith Heyer, "A Survey of Agricultural Development in the Small Scale Farm Areas of Kenya since the 1920's," p. 2, (see note 5) claims Kakamega as the largest labor exporter, as does the 1945 North Kavirondo District *Annual Report*, P. 17.

17 A material consumption scale of 5 points was used to measure economic standing because farmers rarely received a regular wage income. The scale was dichotomized for purposes of simplicity; levels 1 and 2 are low, and levels 3-5, high. At the low economic standing level, farmers had a thatched roof and very little furniture in the house -- not more than a table and two folding chairs plus some cooking pots. At the high level, farmers lived in a permanent or semipermanent house and had more furniture than the basic table and chairs. Permanent houses have concrete walls and floors, as well as corrugated iron sheet roofs. Semipermanent houses have corrugated iron sheet roofs; their mud walls are packed into a frame made of long, wiry branches fastened together, and their mud floors are beaten to become hard and level.

18 This figure is calculated from the 1969 *Census*, Vol. 1, p. 123, by subtracting the number of "excess" women aged 25-29 from the number of men in the same age group. According to the *ILO/UNDP, Employment, Incomes and Equality*, a third of Kenyan households have a male head working away in town (400,000) and another 125,000 are female headed, p. 47.

19 This figure is based on questions and observations during interviews, but especially on a more intensive study of the village in which I lived. Respondents are reluctant to state how much support they receive from distant husbands.

20 Interview in Shigokho, April, 1975. All quotes such as these come from field notes.

21 Achola Pala Okeyo, "Daughters of the Lakes and Rivers," *Women and Colonization*, Mona Etienne and Eleonor Leacock, eds. (New York: Praeger, 1980).

22 For example, at the Kakamega District International Women's Year Seminar, March, 1975.

23 David K. Leonard, "Organizational Structures for Productivity in Agricultural Extension," in his *Rural Administration in Kenya* (Nairobi: East African Literature Bureau, 1973), pp. 134, 145. For a penetrating analysis of the incentive structure in which extension officers work, see Leonard, *Reaching Peasant Farmers: Organization Theory and Practice in Kenya* (Chicago: University of Chicago Press, 1977).

24 Because this is not a random sample and thus does not reflect a normal distribution, The Chi-Square Tests of Significance are technically not appropriate. The size of the sample may mean it approximates a normal distribution, and therefore tests of significance have been included for exploratory purposes. Table size of less than 212 indicate missing data or are subsamples, as specified in the table title. Numbers have been rounded off to the nearest percent.

25 This represents one-sixth of the cost of school fees to send a child to Standards 5, 6, or 7 in a government primary school (1975). The price in 1975 for a bag of hybrid maize seeds -- enough for one acre -- was Shs. 23.

26 In 1975, minimum acreage for the GMR was 15 acres, although I knew of several cases where farmers with 9 acres received loans. The figure comes from the *Development Plan*. Vol. I, p. 212 along with a statement that the government wants to change this disparity.

27 On resource constraints, see Warren F. Ilchman and Norman Uphoff, *Political Economy of Change* (Berkeley: University of California Press, 1971), pp. 32, 58-59: Robert H. Jackson, "Administration and Development in Kenya: A Review of Problems Outstanding," *Development Administration; The Kenyan Experience*, Goran Hyden, Robert Jackson, and John Okumu, eds. (Nairobi: Oxford University Press, 1970), pp. 326; Guy Hunter, *Modernizing Peasant Societies: A Comparative Study in Asia and Africa* (London: Oxford University Press, 1969), pp. 172-73, 187; Albert Waterston, *Development Planning: Lessons from Experience* (Baltimore: Johns Hopkins University Press, 1965). The frequent use of extension-farmer ratios to indicate quality or improvement is also suggestive of this approach.

28 Rogers, 1962, pp. 40, 285-292.

29 The 1975 prices of grade cows started at Shs. 1200/

- or about $200, and local cows were half that price. The price of a grade cow was equivalent to secondary school fees for two children at a government supported school in 1975.

30 The hybrid maize-bean adoption is designated as both a food and cash crop. The cash crops include the hybrid maize-beans combination, coffee, European vegetables (tomatoes, cabbages, or onions), passion fruit, and sugar cane. The food crops include the hybrid maize-beans combination, European vegetables, root crops (sweet potato or cassava), millet crops (finger millet or sorghum), and nut crops (groundnuts or monkey nuts).

31 The cumulative services refer to 1 visit, 2 or more visits, farmer training, demonstration plot attendance, and loan application or receipt. Added to that for Lirhembe members are maize and fertilizer loans, tractor services, and grade cow receipt.

32 A full analysis of this organization can be found in my "The Umoja Federation: Women's Cooptation into a Local Power Structure", *Western Political Quarterly*, 33, 2 July, 1980, pp. 278-290.

33 For the African figure, see Burton Swanson and Jaffar Rossi, *International Directory of National Extension Systems* (Urbana: University of Illinois, 1981), p. 272.

34 Uma Lele, *The Design of Rural Development: Lessons from Africa* (Baltimore: Johns Hopkins University Press, 1975), p. 77.

35 *Development Plan*, P. 18.

36 Approaches like this can be found in Robert Chambers, Managing Rural Development (Uppsala: Scandinavian Institute of African Studies, 1974) and Leonard, 1977.

37 Kenya now requires male agricultural instructors to take a home economics subject during training. Such courses are also offered in in-service training, I am grateful to Janice Jiggins for this information as well as other information about recent changes in Kenya.

38 Kenya is one of the few countries to have instituted a woman-sensitive Training and Visit system which requires the identification of women-contact farmers and women in contact groups. (Again, I am grateful to Janice Jiggins for this point.) However, although easily adaptable, most Training and Visit systems still perpetuate male preferential patterns. For information on Training and Visit systems see *Agricultural Extension: Training and Visit System*, by Daniel Benor and James Q. Harrison, and *A Systems for Monitoring and Evaluating Agricultural Extension Projects*, by Michael Cernea and Benjamin Tapping (both,

Washington DC, The World Bank, 1977).

39 This "people-oriented" approach as opposed to top-down approaches in outlined more fully in Felipe Alfonso and David Karten, *Bureaucracy & The Poor* (West Hartford, Kumarian Press, 1980) and David Korten, "Community Organization and Rural Development: A Learning Process Approach" *Public Administration Review* Vol. 40 Sept-Oct 1980 pp. 400-510

I recoginize that most development agencies & donor organizations currently operate in ways incompatible with this approach. However, some private voluntary organizations have used this approach successfully.

# APPENDIX

## SAMPLING AND INTERVIEWING PROCEDURES

A geographically purposive sample of 212 farm households was obtained in the Idakho research site between December, 1974, and June, 1975. It represents 10 percent of the total number of households in the geographic areas targeted. My initial concern was to assure that varying distances from the road and main paths, and thus from agricultural instructors and services, would be covered. These geographic areas coincide with clan and subclan identities. Once spatial areas were designated in order to obtain geographic and clan representativeness, I selected farms that would be representative of varying economic standings and age groupings. Numbers were based on my approximations of their proportion of the population. I did not know in advance, however, about the farm management type until the interview had begun. The correspondence of female managers to the proportion of female heads in Kakamega supports the notion that my choice of farmers was chance-like in method. The sample is not, however, a random one, and the universe of this sample is restricted to one location. The sample does not purport to generalize to all of Kenya or Africa, but rather to illustrate sex differences within a sample which may be suggestive for other parts of Kenya or Africa with agriculturally based cultures in areas of high land scarcity and rates of male out-migration.

Though scientific sampling techniques were not utilized, I am confident that the sample judiciously represents a reasonable cross section of farmers in high and ordinarily serviced areas in western Kenya. The basis of this confidence is my six-month residence in Shikulu and my participation in community life, where I gained in-depth knowledge of that subclan and geographic area.

A female research assistant from Shikulu and I conducted the interviews. My assistant translated questions and responses from Lwidakho to English. The interview format generally proceeded as follows. We entered a farm compound, greeted household members, introduced ourselves, and requested an interview. (No respondents declined to be interviewed.) We then proceeded to ask a systematic set of questions of each farmer, beginning with demographic and crop information, land size, husbandry practices, and sources of information about crops and practices. Farmers live in dispersed settlement patterns, and their fields surround houses. This afforded us the opportunity of visually verifying answers to questions relating to land size and crop husbandry practices. The next section of the interview involved questioning farmers about their awareness, receipt, and evaluation of agricultural services. Following that, we asked questions about community participation and electoral activity. For female respondents, questions were then asked which were designed to tap attitudes toward women and women's problems. Probing about marital situations depended greatly on the willingness of respondents to discuss such matters. No systematic responses of personal marital relations have been tabulated because of the unevenness of answers.

Great efforts were made to make the interview as informal, conversational, and reciprocal as possible. No questionnaire was utilized; instead, notes were taken during the interview, which I then typed at the end of the day. When the interview was completed, respondents were invited to interview and probe the interviewer. This exchange permitted deeper insight into the respondent and a more holistic understanding of the farm. The length of the interview depended on the openness of respondents. A typical interview took 40 minutes; minimal length interviews took 20 minutes, and the lengthiest were 3 hours.

# BIBLIOGRAPHY


Ascroft, Joseph; Chege, Fred; Kariuki, Joseph; Roling, Nieles; and Ruigu, George. "Does Extension Create Poverty in Kenya?" *East Africa Journal*, 1972.

Benor, Daniel, and Harrison, James Q. *Agricultural Extension: Training and Visit System*. Washington, D.C.: The World Bank, 1977.

Boserup, Ester. *Women's Role in Economic Development*. New York: St. Martin's Press, 1970.

Bowe, Edgar, and Moris, Jon. "Social Characteristics of Progressive Buganda Farmers." *East African Journal of Rural Development* 2, 1, (1969).

Bukhe, Jette. *Village Women in Ghana*. Uppsala: Scandinavian Institute of African Studies, 1979.

Cernea, Michael, and Tepping, Benjamin. *A System for Monitoring and Evaluating Agricultural Extension Projects*. Washington, D.C.: The World Bank, 1977.

Chambers, Robert. *Managing Rural Development*. Uppsala: Scandinavian Institute of African Studies, 1974.

de Wilde, John. *Experiences with Agricultural Development in Tropical Africa*. Baltimore: Johns Hopkins University Press, 1967.

Fortman, Louise. "Women's Work in a Communal Setting: The Tanzanian Policy of Ujamaa." In *Women and Work in Africa*, edited by Edna Bay. Boulder: Westview Press, 1982.

Hailey, Lord. *An African Survey*. London: Oxford University Press, 1956.

Heyer, Judith. *A Survey of Agricultural Development in Small Scale Farm Areas of Kenya Since the 1920's*. Working Paper no. 194. Nairobi: University of Nairobi, Institute for Development Studies, 1974.

Heyer, Judith, and Ascroft, Joseph. "The Adoption of Modern Practises on Farms in Kenya: Preliminary Results of a 1968 Survey of Farms Across Kenya." In *Documents of the East Africa Social Science Conference*. Dar es Salaam: 1970.

Heyer, Judith; Ireri, D.; and Moris, Jon. *Rural Development in Kenya*. Nairobi: East African Publishing House, 1971.

Hunter, Guy. *Modernizing Peasant Societies: A Comparative Study in Asia and Africa.* London: Oxford University Press, 1969.

ILO/UNDP. *Employment, Incomes and Equality: A Strategy for Increasing Employment in Kenya.* Geneva: ILO, 1972.

Ilchman, Warren F., and Uphoff, Norman. *Political Economy of Change.* Berkeley: University of California Press, 1971.

Jackson, Robert H. "Administration and Development in Kenya: A Review of Problems Outstanding." In *Development Administration: The Kenyan Experience,* edited by Goran Hyden, Robert Jackson, and John Okumu. Nairobi: Oxford University Press, 1970.

Kenya, Republic of. *African Socialism and its Application to Planning in Kenya.* Sessional Paper no. 10, 1965.

______, _________ __. Agricultural Education Commission. *Report of the Agricultural Education Commission.* Nairobi: Government Printer, 1967.

______, _________ __. Ministry for Finance and Economic Planning. *Development Plan for the Period 1974 to 1978.* Nairobi: Government Printer, 1974.

______ _________ __. Ministry for Finance and Economic Planning. *Kenya Population Census 1969,* Nairobi: 1970.

______, _________ __. Ministry of Economic Planning and Development. *A Report on Economic Studies of Farming in Nyanza Province, 1963.* Farm Economic Report no. 26.

Kenya Colony and Protectorate. Nyanza Province. North Kavirondo District. *Annual Report,* 1948. Nairobi, Kenya National Archives, Native Agricultural File.

_____ _______ ___ ______________. *Annual Report, 1951.* ________, _____ _________ _________, _____ ______________ _____.

_____ _______ ___ ______________. *Annual Report, 1952.* ________, _____ _________ _________, _______ ______________ _____.

_____ _______ ___ ______________. _______ _________. _____ _________ _________. *Annual Report, 1959.* _________, _____ _________ _________, _______ ______________ _____.

_____ _______ ___ ______________. _______ _________. _____ ________. _____ __________ _________. *Annual Report, 1955.* "Special Supplement." Nairobi, Kenya National Archives, Native Agricultural File.

______ ______ ___ ____________. *A Plan to Intensify the Development of African Agriculture in Kenya*. R.J.M. Swynnerton, comp. Nairobi: Government Printer, 1954.

Korten, David C. "Community Organization and Rural Development: A Learning Process Approach." *Public Administration Review* (Sept.-Oct. 1980).

Korten, David C., and Alfonso, Felipe. *Bureaucracy and the Poor*. West Hartford: Kumarian Press, 1980.

Lele, Uma. *The Design of Rural Development: Lessons from Africa*. Baltimore: Johns Hopkins University Press, 1975.

Leonard, David K. "Organizational Structures for Productivity in Agricultural Extension." In his *Rural Administration in Kenya*. Nairobi: East African Literature Bureau, 1973.

________, ______ _. *Reaching Peasant Farmers: Organization Theory and Practice in Kenya*. Chicago: University of Chicago Press, 1977.

________, _______ _. "Why do Kenya's Agricultural Extension Services Favor the Rich Farmers?" Paper presented at the 16th Annual Meeting of the African Studies Association, Syracuse, October-November 1974.

Moock, Peter. "The Efficiency of Women as Farm Managers: Kenya." *American Journal of Agricultural Economics* 58, 5 (December 1976).

______, ______. *Managerial Ability in Small Farm Production: An Analysis of Maize Yields in the Vihiga Division of Kenya*. Ph. D. diss., Columbia University, 1973.

Nairobi. Kenya National Archives. Native Agricultural File, Center and North Kavirondo (1930-46).

Okeyo, Achola Pala. "Daughters of the Lakes and Rivers." In *Women and Colonization*, edited by Mona Etienne and Eleonor Leacock. New York: Praeger, 1980.

Ominde, S.H. *Land and Population Movements in Kenya*. Nairobi: Heineman, 1968.

Rogers, Everett. *The Diffusion of Innovations*. New York: Free Press, 1962.

_______, ________. *Modernization Among Peasants: The Impact of Communication*. New York: Free Press, 1969.

Sangree, Walter. *Age, Prayer and Politics in Tiriki, Kenya*.

London: Oxford University Press, 1966.

Sheffield, James, ed. *Education, Employment and Rural Development*. Nairobi: East African Publishing House, 1967.

Strange, Bertha F. *The Story of Kenya's Home Extension Service: 1962-65*. Washington, D.C.: U.S.A.I.D. Mission to Kenya, 1965.

Staudt, Kathleen. "The Umoja Federation: Women's Cooptation into a Local Power Structure." *Western Political Quarterly* 33, 2 (July 1980).

Swanson, Burton, and Rossi, Jaffar. *International Directory of National Extension Systems*. Urbana: University of Illinois, 1981.

Wagner, Gunter. *The Bantu of North Kavirondo*. 2 vols. London: Oxford University Press, 1949.

Waterston, Albert. *Development Planning: Lessons from Experience*. Baltimore: Johns Hopkins University Press, 1965.